# Geographic Information Systems for Group Decision Making

T0330488

# Geographic Information Systems for Group Decision Making

Towards a participatory, geographic information science

Piotr Jankowski and Timothy Nyerges

London and New York

Published 2001 by Taylor & Francis
2 Park Square, Milton Park, Abingdon, Oxon OX14 4RN
52 Vanderbilt Avenue, New York, NY 10017

First issued in paperback 2020

*Taylor & Francis is an imprint of the Taylor & Francis Group, an informa business*

Typeset in Times by Wearset, Boldon, Tyne and Wear

*British Library Cataloguing in Publication Data*
A catalogue record for this book is available from the British Library

*Library of Congress Cataloging in Publication Data*
A catalogue record for this book has been requested

ISBN 13: 978-0-367-57884-8 (pbk)
ISBN 13: 978-0-7484-0932-7 (hbk)

# Contents

# Series introduction

## Welcome

The *Research Monographs in Geographical Information Systems* series provides a publication outlet for research of the highest quality in GIS, which is longer than would normally be acceptable for publication in a journal. The series includes single- and multiple-author research monographs, often based upon PhD theses and the like, and special collections of thematic papers.

## The need

We believe that there is a need, from the point of view of both readers (researchers and practitioners) and authors, for longer treatments of subjects related to GIS than are widely available currently. We feel that the value of much research is actually devalued by being broken up into separate articles for publication in journals. At the same time, we realise that many career decisions are based on publication records, and that peer review plays an important part in that process. Therefore a named editorial board supports the series, and advice is sought from them on all submissions.

Successful submissions will focus on a single theme of interest to the GIS community, and treat it in depth, giving full proofs, methodological procedures or code where appropriate to help the reader appreciate the utility of the work in the Monograph. No area of interest in GIS is excluded, although material should demonstrably advance thinking and understanding in spatial information science. Theoretical, technical and application-oriented approaches are all welcomed.

## The medium

In the first instance the majority of Monographs will be in the form of a traditional textbook, but, in a changing world of publishing, we actively encourage publication on CD-ROM, the placing of supporting material on web sites, or publication of programs and of data. No form of dissemunation is discounted, and prospective authors are invited to suggest whatever form of publication and support material they think is appropriate.

## The editorial board

The Monograph series is supported by an editorial board. Every monograph proposal is sent to all members of the board which includes Ralf Bill, António Câmera, Joseph Ferreira, Pip Forer, Andrew Frank, Gail Kucera, Peter van Oostrom, and Enrico Puppo. These people have been invited for their experience in the field, of monograph writing, and for their geographic and subject diversity. Members may also be involved later in the process with particular monographs.

## Future submissions

Anyone who is interested in preparing a Research Monograph should contact either of the editors. Advice on how to proceed will be available from them, and is treated on a case by case basis.

For now we hope that you find this, the sixth in the series, a worthwhile addition to your GIS bookshelf, and that you may be inspired to submit a proposal too.

Editors:

Professor Peter Fisher
Department of Geography
University of Leicester
Leicester
LE1 7RH
UK
Phone: +44 (0) 116 252 3839
Fax: +44 (0) 116 252 3854
Email: pff1@le.ac.uk

Professor Jonathan Raper
Schools of Informatics
City University
Northampton Square
London
UK
Phone: +44 (0) 20 7477 8000
Fax: +44 (0) 20 7477 8587
Email: raper@soi.city.ac.uk

# Preface

Groups are said to be the basic building blocks of society. They mediate interests and help give voice to social, health, environmental, economic, and safety concerns, to name but a few. Decision making within both small and large groups is perhaps one of the more important activities of group behavior. Decision making establishes direction for action. Within the private sector over the past 15 years, organizational development has followed a trend toward flatter structures. That means more participation in the direction of what and how things are accomplished in an organization. Within the public sector, citizen participation grows in significance as more citizens claim ineffective political representation on the part of elected officials about placed-based public decision problems. Within the link between the private and public sectors over the last several years there has been a trend of private industry working more closely with public organizations, in so called private–public coalitions, to explore win-win situations for solving difficult community problems. In a similar manner, the rise of non-governmental organizations is in some way due to the ineffectiveness of governments to respond to the needs and call for action, and the shortcomings of private industry in pursuing a narrow, capitalistic motivation — called "profits" — when coming to grips with various valued concerns. Stove-piping of decision activities, whereby only one perspective is given voice for a long time in the private and/or public sector, might have caused many of the problems currently facing communities throughout the world. The complexity of many public–private situations is thus brought about by "stove-pipe responsibility", hence lack of accountability for those who have or who have not acted. Communities, whether place-based or cyber-based, are ripe for political restructuring. The growth in group decision making activity in essence is a restructuring of the political scene on the local, regional, national, and international scales. Of course, the fuel added to the fire of restructuring change depends on the particular situations from place to place and what kinds of information are available.

One of the fundamental freedoms in a democratic society is the right of a citizen to know and participate in a decision situation, when decisions about valued-concerns are being made that affect the welfare (taken broadly) of those people and the places they live in. This is particularly true when those situations involve public or public–private problems, and the impacts occur to community on a local, state, regional, national, and global scale. It seems that representative democracy is being challenged in a way by modern communications technology. With direct access to information communication technology comes an impression that direct democracy is better due to closer ties to information. The internet is at the core of a change in getting access to information in a timely manner. Access to wireless internet communications technology — which is on the verge of a substantial expansion — will be likely to fuel the frustration in decision situations. The continual lament is: Why isn't more being done faster?

Getting access to information about valued-concerns in community and society is one of the reasons why geographic information systems (GIS) are being put to use, but certainly not the only reason. Through broader access to GIS data it is expected that people can analyze and deliberate the pros and cons of values, goals, objectives, and criteria describing public and public–private problems at various scales. Whether this slows or improves any given decision situation, and decision situations in general, still remains to be seen. Nonetheless, more and more information is being made available for groups and citizens to consider if they so choose. Creating an environment to facilitate analysis and deliberation in a group decision setting is the purpose behind participatory GIS (PGIS). Developing a conceptual understanding of the use of PGIS, which in turn might add to a more effective deployment of PGIS, as one among many viable information technologies, is the purpose behind this book.

This book has been written as an equal effort between the co-authors. It is a report of research activities between 1995–2000. Although much of our research activity related to this topic has been published in journals in one form or another, the book contains eight original chapters as a synthesis of findings. Researching the dynamics of complex geographic decision situations, examining the influences of the use of participatory geographic information systems and its extension as a form of decision support capability, is the principle motivation for undertaking the investigations reported herein. We see the research as forming a foundation for what we call "participatory, geographic information science".

This book is meant to be an introduction to participatory, geographic information science as much as it is a report on our research agenda for the past few years. The foundation of this book is built from a concerted effort to balance three research domains — theory, methodology, and substance — involved in studies of PGIS use. All three domains are (or rather should be) present in all research, but the difference in research is a matter of the difference in emphasis of the domains as used in a research study. We have tried to make this clearer by writing this book in order to open opportunities for research, not stifle them.

The book is structured as follows. In Chapter 1 we set the tone about how these three research domains can be combined to set the research orientation of a study. Understanding the balance of emphasis among domains leads one to understand the difference in research orientation as basic, method-driven, and applied research. Understanding the difference in emphasis as to which domain leads the emphasis, which domain supports, and which domain follows, sets up a "pathway" as the basis of research strategies reported in the three empirical studies reported herein. Much of this book is about the conceptual underpinnings of participatory decision making. We treat these issues in Chapter 2, in the form of Enhanced Adaptive Structuration Theory 2 that relates the convening, process, and outcome aspects of decision situations to each other within the context of a human-computer-human interaction. In regards to methodology, we are not afraid of being labeled methodologists, from the perspective of both GIS decision support methods and social-behavioral methods as they are treated in Chapters 3 and 4, respectively. Chapter 3 highlights the methods and tools that underpin PGIS as an extended set of capabilities to standard GIS capabilities. In Chapter 4 we provide a comprehensive overview of how research strategies can be designed to investigate PGIS use in participatory decision making. Those chapters set the stage for the chapters of the second part of the book.

In Chapters 5, 6, and 7 we present three studies that address substantive decision making concerns about public health, transportation, and habitat restoration, respectively. We made use of three rather different research strategies to develop empirical findings. Each of the findings stems from the emphasis of the three domains. Chapter 5 treats public health decision making as a problem in task analysis to elucidate the character of geographic decision support capabilities. Chapter 6 uses a case analysis approach to investigate a transportation improvement program decision process to uncover the influence among a variety of decision aspects and speculates about why GIS is not used more often in such situations. Chapter 7 reports on a group experiment concerning habitat restoration, in which the data that resulted from the experiment are analyzed using two different approaches, and the approaches are compared in terms of the amount of information gain each provides to the findings. In the conclusions of Chapter 8 we reflect on how the emphasis of the three domains was used, and what prospects there are for future research.

Given the trends involving the growth of participation in public-private decision making and the trends in technology change, we see a tremendous opportunity for research in participatory geographic information systems development and use. Through a better understanding among three research domains, and how each supports and at the same time constrains each other, we hope that this book will motivate the reader to make a contribution in some manner toward a participatory, geographic information science.

Piotr Jankowski
Timothy Nyerges

# Acknowledgements

Much of the material presented in this book is the outgrowth of the authors' research, funded in part by National Science Foundation programs in Geography and Regional Science and Decision, Risk and Management Science under Grant No. SBR-9411021, the Idaho State Board of Education, and the Idaho Department of Health and Welfare. The above support does not constitute an endorsement by the organizations of the views expressed herein, but we are very grateful for the support.

In developing the material for this book we benefited from interactions with our colleagues: Gennady and Natalia Andrienko, Thomas Gordon, Hans Voss and Claus Rinner from the German National Research Center of Information Technology in Sankt Augustin, Alexander Lotov from the Computing Center of the Russian Academy of Sciences, Robert McMaster of the University of Minnesota, and Kerry Brooks of Clemson University. We thank Jay Clark and Karen Richter of the Puget Sound Regional Council for providing interview time and the materials about transportation improvement programs of our review and analysis. Our thanks also go to Jane Smith and Chris Maddock of the Idaho Department of Health and Welfare for helping us study the decision process of allocating funds to primary health care. We acknowledge the helpful input of members of the Elliott Bay/Duwamish Restoration Panel, especially Robert Clark, Jr., and Jennifer Stegner of NOAA Restoration Northwest, Curtis Tanner of the US Fish and Wildlife Service, Glen St. Amant of the Muckleshoot Tribe, and Bob Matsuda of King County Department of Metropolitan Services, who provided a critique of the Spatial Group Choice software at an early stage of development. We thank the personnel associated with Taylor & Francis, particularly Tony Moore and Edith Henry, who made this volume possible by taking care of details. In combination with that acknowledgement, we thank the GIS research monograph series editors Drs Peter Fisher and Jonathan Raper for understanding the significance of this topic.

Special thanks go to our former graduate students: T.J. Moore, Milosz Stasik, Alan Smith, Robb Montejano, and Marcie Compton who helped tremendously as colleagues and research assistants in various research projects of which the results are reported in this book in part or in their entirety. We thank our current graduate students Christina Drew, Charles Hendricksen, Nicholas Hedley and David Tuthill for their interest in listening and for the courage to add their critique to what we had to say. Last but not least, we are indebted to our families. Without their love, support and patience, the writing of this book would not have been possible.

# 1 Introduction to geographic information systems and participatory geographic information science

**Abstract**

Group decision making that deals with geographic problems has been around for quite some time. However, an interest in participatory decision making is growing in importance as more and more people with concerns about environmental, land use, natural resource, and transportation issues believe that those who are affected by decisions should be a part of the process. Many geographic decision problems are viewed as unstructured and laden with locational conflict because their solutions demand the participation of multiple stakeholders with varying stakeholder values. In this introductory chapter we introduce the reader to what we call "participatory geographic information systems" (PGIS) and provide an overview of what we call "participatory geographic information science". Geographic information systems that are designed and used by groups with multiple stakeholder perspectives are described as "participatory geographic information systems". Participatory geographic information systems have all of the capabilities of GIS, with additional capabilities for group decision support. Social-behavioral studies about the use of participatory geographic information systems as a process of human-computer-human interaction are a cornerstone of the empirical aspect of participatory geographic information science. Participatory geographic information science is a subfield of geographic information science that contributes to an understanding of PGIS use in society. We introduce the reader to our framework for this book that is based on balancing the emphasis among research domains — theory, method, and substance. That framework underpins our approach to research and helps us build toward a participatory geographic information science.

Spatial decision making is an everyday activity, common to individuals and organizations. People make decisions influenced by geography when they choose a store to shop, a route to drive, a path to jog, or a neighborhood for a place to live, to name but a few. Organizations are not much different in this respect. They take into account the realities of spatial organization when selecting a site, choosing a land development strategy, allocating resources for public health, and managing infrastructures for transportation or public utilities.

Most of the individual spatial decisions are made ad hoc, without a formal analysis. Such decisions are often based on experiential heuristics and internalized preferences (values). This expedient approach to spatial decision making can sometimes be explained by a relatively small "decision equity" at stake in daily decision situations, such as the selection of a place to shop or an entertainment venue. The cost of making a poor choice (decision) can be a smaller selection of goods, higher prices paid than elsewhere, or a boring evening spent at a movie theater. In contrast to these everyday decision situations faced by individuals, the decision equity for organizations and inter-organizational coalitions is often quite high. Consequently, organizations and inter-organizational coalitions are more likely to use an analytical approach to support the decision making process.

Current trends in modern organizations towards flatter structures and the involvement of many stakeholder groups in solving decision problems have created a need for information technology capable of supporting participatory decision making. Such information technology has developed in recent years for the computerized support of group decision making which is aimed at solving business problems such as market strategies, corporate planning, and product development. Group decision support solutions are now offered as commercial products, developed on the premise that workgroups will dominate the emerging organizational structures of the near future (Orsborne *et al.* 1990, Coleman and Khanna 1995).

Similar information technology aimed at solving spatial decision problems, e.g. land use/resource development negotiations, site selection, choice of environmental and economic strategies, and urban/regional development, is now being discussed in geography and planning literature (Armstrong 1993, Faber *et al.* 1995, Nyerges 1995, Shiffer 1992, 1995). This surge of interest in collaborative spatial decision making (CSDM) in particular, and participatory decision making more generally, has been spurred not only by the trend in business organizations, but foremost by the realization that effective solutions to spatial decision problems require collaboration and consensus building. Many spatial problems are labeled as "wicked" or difficult (Rittel and Webber 1973) because they contain intangibles that cannot be easily quantified and modeled, their structure is only partially known or burdened by uncertainties, and potential solutions often become locally unwanted land uses (LULUs) that instigate not in my back yard (NIMBY) controversies. These include landfill and hazardous waste facility siting (Couclelis and Monmonier 1995, Lake 1987), polluted urban land use (so-called brownfield) redevelopment projects (Davis and Margolis 1997, Bartsch and Collaton 1997), and salmon habitat restoration plans (NOAA 1993, Brunell 1999). These problems require the participation and collaboration of people representing diverse areas of competence, political agendas, and social interests. As a consequence, diverse groups must often be involved to generate solutions to pervasive spatial problems (Golay and Nyerges 1995).

The need for computerized decision support results from the importance of group decision making and problem solving carried out predominantly during meetings, and from common problems associated with meetings such as: overemphasis on social-emotional rather than task activities, failure to define a problem adequately before rushing to judgement, pressure, inhibiting creativity, felt by subordinates in the presence of bosses, and the feeling of disconnection/alienation from the meeting (Nunamaker *et al.* 1993). A number of other problems hampering the effectiveness of meetings is given by Mosvick and Nelson (1987) and include (after Lewis 1994): getting off the subject, too lengthy, inconclusive, disorganized, no goals or agenda, dominating individuals, not effective for making decisions, rambling, redundant, or digressive discussion. Despite these negative characteristics, the attractiveness of a group approach to decision making comes in general from the fact that individual contributions are increased by a synergistic effect resulting from meeting dynamics. Sage (1991) identifies several human decision making abilities that information technology might augment in meetings. These include:

1   help decision makers formulate, frame, or assess decision situations by identifying the salient features of the environment, recognizing needs, identifying appropriate objectives by which to measure the successful resolution of an issue;
2   provide support in enhancing the abilities of decision makers to obtain and analyze possible impacts of alternative courses of action; and
3   enhance the ability of decision makers to interpret impacts in terms of objectives, leading to an evaluation of alternatives and selection of a preferred alternative option.

Consequently, a final outcome of a computer-supported decision meeting can be more than a simple sum of individual contributions. The attractiveness of a computer-supported group approach to spatial decision making comes from a possibility of engaging diverse participants as competent stakeholders through computer-mediated communication, problem exploration, and negotiation support. An information technology that can potentially be exploited to facilitate computer-supported approach to spatial decision making is geographic information systems (GIS).

GIS use has expanded in society in the last decade faster than any other analytical information technology. New developments of the 1990s, focused on the internet and the world wide web, opened new possibilities for better access to spatial information and enhanced benefits from its use.

While the mainstream GIS technology concentrated on the creation of easy-to-use, ubiquitous mapping and spatial analysis tools, it lacked a capability to collate interests and interactions to support collaborative spatial decision making (CSDM), for example, in the context of face-to-face meetings. This and other capabilities (e.g. supporting collaborative work distributed in space and time) are needed to enhance widespread citizen participation in public spatial decision making, such as land use planning, and to bring to a fuller realization the democratic maxim that those affected by a decision should participate directly in the decision making process.

Participatory land use planning involving citizens is not the only type of collaborative spatial decision making. Resource development and environmental management figure prominently on the list of spatial decision problems that can benefit from a group/collaborative approach. Examples of these problems, included in a review conducted by the US President's Council on Environmental Quality (CEQ), are the Glen Canyon Dam project and the Ozark Mountain Highroad project (CEQ 1997). The first project dealt with developing a management plan to regulate the operation of Glen Canyon Dam located in Arizona. How the dam was going to be operated would have significant impacts on the Grand Canyon and other downstream resources. The project involved the participation of five federal agencies, one state agency, six Native American tribes and a large number of citizens (over 33,000 commented on the draft environmental impact statement). A GIS was used to manipulate and map project-critical information so that collaborating parties could understand the data and develop decision alternatives. The Ozark Mountain Highroad project dealt with traffic congestion in one of the popular entertainment centers in the USA — Branson, Missouri. Famous for its country music, Branson attracts tens of thousands of motorized tourists congesting Country Music Boulevard on a daily basis, causing huge traffic delays. Challenged by the governor of Missouri, the state Highway and Transportation Department embarked in 1992 upon a collaborative planning project to develop a new four-lane highway in six months without neglecting environmental integrity. The project involved local, state, and federal agencies and consulting firms, and resulted in the development of a number of feasible design alternatives providing the basis for decision making.

Spatial decision making problems commonly involve three categories of participants: stakeholders, decision makers, and technical specialists. The diversity of participant categories may include a range of expertise levels in virtually any decision problem — from novice through intermediate to expert. Reducing the complexity of a decision problem by reducing the cognitive workload of participants is one goal of developing collaborative decision support systems. Reducing cognitive workload will hopefully lead to a more thorough treatment of information, by exposing initial assumptions more clearly, facilitating critiques of the accuracy of information, and subsequently resulting in more effective and equitable participatory decisions.

During the 1990s, geographic information systems (Faber *et al.* 1994, Godschalk *et al.* 1992), their offspring spatial decision support systems (SDSS) (Armstrong 1993, Densham 1991), and spatial understanding (and decision) support systems (SUSS/SUDSS) (Couclelis and Monmonier 1995, Jankowski and Stasik 1997) were suggested as information technology aids to facilitate geographic problem understanding and decision making for groups, including groups embroiled in locational conflict. We have chosen to group all of the above mentioned technologies for group GIS under the umbrella term "participatory GIS" (PGIS), a term first used by Harris *et al.* (1995). Geographic information systems that are designed and used by groups with multiple stakeholder perspectives are described as "participatory geographic information systems". In this book we extend the idea of PGIS to include not only the capabilities of GIS, but also additional capabilities for group decision support, e.g. group communication and decision analysis capabilities. Clearly, research about PGIS and collaborative decision making for geographically oriented, public policy problems continues to gain momentum. Unfortunately, most of the research concerning collaborative spatial decision making has been about GIS development rather than about GIS use, without a strong theoretical link between the two. Little has been done until recently to study the use of GIS technology at a decision group level. Even though the case can be made for transferability of research results from experiments with group support systems carried out in the management and decision sciences since the early 1980s, unlike a business decision problem, such as the selection of a product marketing plan, spatial decision problems are unique in making location and associated spatial relationships an explicit part of a decision situation. This gap between the understanding of the implications of using decision support software in non-spatial versus spatial group decision processes is part of the motivation for this book. It is further motivated by the need to develop an understanding of how GIS software combined with decision support techniques is used in group decision processes, which components of computer technology fulfil decision support tasks, and which do not. We believe that this knowledge is needed to enable a better understanding how GIS can be successfully used to support collaborative work involving spatial decision making and problem solving. Such knowledge is needed as participatory and collaborative decision making and problem solving continue to play an ever-increasing role in private–public decision situations of the future.

Our framework for this book is based on balancing the emphasis among research domains — theory, method, and substance — as suggested in social-behavioral research by Brinberg and McGrath (1985). All three domains contribute to research that forms the basis of participatory geographic information science. Participatory geographic information science is a subfield of geographic information science that includes strong linkages among substance, theory, and methods when researching implications of the use of participatory geographic information systems in society. Research about the *development* of participatory geographic information systems has an applied orientation that stems from a leading emphasis within the methodological domain. Research about the *use* of participatory geographic information systems can have an applied or basic orientation depending on whether the lead domain is substantive or conceptual, respectively. Thus, the issue is not whether GIS is tool or science, for this is only a partial perspective (Fisher 1998). The issue really concerns the balance of the domains — substance, theory, and method — that contribute to findings and knowledge about both tools and science, as tools are embedded within research studies as part of participatory geographic information science. In the material presented in this book we are using social-behavioral research methods to examine people's use of GIS methods. Consequently, GIS methods become part of the "substantive domain", just as people, organizations, decision tasks etc, are part of the substantive domain. In our pursuit of social-behavioral methods to examine people's use of GIS we are guided by theories of human-computer-human interaction. These theories are generic in how they treat the use of computer technology in participatory problem solving and decision making. As the consequence of this, we strive to develop a theory that helps us explain how GIS technology is used in participatory problem solving and decision making. Knowledge derived from this effort helps us not only understand how and in what situations GIS technology can be successfully used in participatory decision making, but also develop better GIS methods and tools that will promote such participation.

This book has two goals. The first goal is to broaden and deepen the conceptual underpinnings of participatory GIS by considering social-behavioral aspects of geographic information use. This is a social and behavioral perspective on geographic information science that concerns itself with the implications of geographic information (system) use within groups, organizations, communities, and society. The second goal is to present readers with methods, techniques, and examples of studying and introducing a collaborative approach to spatial decision problem solving. Thus, the book may be treated as a methodological aid for researchers and students interested in the subject of collaborative spatial decision making and as a guide for practitioners interested in introducing a collaborative approach to solving realistic spatial decision problems.

To achieve those goals, the book is organized into two parts that stem from balancing theory, method, and substance. The first part provides theoretical and methodological foundations for participatory spatial decision making. Chapter 2 introduces a conceptual framework for participatory spatial decision making. The chapter discusses multiple perspectives on the importance of communication, cooperation, coordination and collaboration in spatial decision making, including functional, tool, and organizational perspectives, and presents a theoretical framework for studying, analyzing and implementing participatory spatial decision making in practice. Chapter 3 deals with methodological foundations of participatory spatial decision support capabilities. The objective of this chapter is to provide an overview of the methods and technologies that have a promise and potential for participatory GIS. The methodologies discussed include spatial data management, visualization, multiple criteria decision analysis, and mediation and consensus building. Chapter 4, which closes the theoretical/methodological part of the book, focuses on social-behavioral research strategies for studying the use of participatory geographic information systems. The chapter presents a systematic treatment of strategies in terms of the level of induced control in social-behavioral relations in a research setting and the amount of pre- or post-structuring of data for data collection. The objective of this chapter is to provide guidelines for researchers interested in understanding the advantages and disadvantages of choosing various strategies for studying group use of participatory GIS.

In the second part, the theoretical/methodological foundations of PGIS are combined with substantive domains represented by realistic decision scenarios at three different scales. Chapter 5 presents a decision scenario involving primary health care funding for the state of Idaho in the northwest United States. Chapter 6 presents a decision scenario concerning transportation improvement program decision making in a regional transportation planning context in the central Puget Sound Region of Washington State. Chapter 7 presents a decision scenario concerning habitat restoration along a waterway channel in the greater Seattle area of central Puget Sound. In each of those chapters we forge a link between a substantive domain (the decision scenarios), and the theory and methods of previous chapters, thus demonstrating how three domains (theory, methods, and substance) can be balanced in research concerning participatory geographic information science. We use a theory to guide us in our empirical research investigations of substantive spatial decision situations involving the methods of GIS-supported collaborative decision making. We conclude the book with Chapter 8 which offers our perspective on prospects and future directions for research about GIS and group decision making.

## References

Armstrong, M.P. (1993) "Perspectives on the development of group decision support systems for locational problem solving", *Geographical Systems*, 1(1): 69–81.

Bartsch, C. and Collaton, E. (1997) *Brownfields: Cleaning and Reusing Contaminated Property*, Westport, Praeger, Conn.

Brinberg, D. and McGrath, J.E. (1985) *Validity and the Research Process*, Thousand Oaks, Sage.

Brunell, D. (1999) "Not just any salmon plan will do the job", *Eastside Business Journal*, Bellevue, Business Journal Publications: 23.

CEQ (1997) "National Environmental Policy Act. A study of its effectiveness after twenty-five years", Council on Environmental Quality, Executive Office of the President. Available at: http://ceq.eh.doe.gov/nepa/nepanet.htm. Accessed on 4/25/00.

Coleman, D. and Khanna, R. (1995) *Groupware: Technologies and Applications*, Upper Saddle River, NJ, Prentice Hall.

Couclelis, H. and Monmonier, M. (1995) "Using SUSS to resolve NIMBY: how spatial understanding support systems can help with the 'not in my back yard' syndrome", *Geographical Systems*, 2(2): 83–101.

Davis, T. and Margolis, K. (eds) (1997) *Brownfields: A Comprehensive Guide to Redeveloping Contaminated Property*, Chicago, American Bar Association.

Densham, P.J. (1991) "Spatial decision support systems", in D.J. Maguire, M.F. Goodchild and D.W. Rhind (eds) *Geographical Information Systems: Principles and Applications*, New York, John Wiley & Sons: 403–12.

Faber, B., Knutson, J., Watts, R., Wallace, W., Hautalouma, J. and Wallace, L. (1994) "A groupware-enabled GIS", *GIS '94 Symposium*: 551–61.

Faber, B., Wallace, W. and Cuthbertson, J. (1995) "Advances in collaborative GIS for land resource negotiation", *Proceedings, GIS '95*, Ninth Annual Symposium on Geographic Information Systems, Vancouver, BC, March 1995, GIS World, Inc., Vol. 1: 183–89.

Fisher, P. (1998) "Geographic Information Science and Systems: a Case of the Wrong Metaphor", in T.K. Poiker and N.R. Chrisman (eds) *Proceedings*, 8th International Symposium on Spatial Data Handling, Simon Fraser University, Department of Geography, International Geographical Union, Geographic Information Science Study Group: 321–30.

Godschalk, D.R., McMahon, G., Kaplan, A. and Qin, W. (1992) "Using GIS for computer-assisted dispute resolution", *Photogrammetric Engineering and Remote Sensing* 58(8): 1209–12.

Golay, F. and Nyerges, T. (1995) "Understanding collaborative use of GIS through social cognition", in T. Nyerges, D.M. Mark, R. Laurini and M. Egenhofer (eds), *Cognitive Aspects of Human-Computer Interaction for Geographic Information Systems*, Proceedings of the NATO ARW, Mallorca, Spain, March 21–25, 1994. Dordrecht, Netherlands, Kluwer Academic Publishers: 287–94.

Harris, T.M., Weiner, D., Warner, T. and Levin, R. (1995) "Pursuing social goals through participatory GIS: redressing South Africa's historical political ecology", in Pickles, J. (ed.) *Ground Truth: The social implications of geographical information systems*, New York, NY, Guilford: 196–222.

Jankowski, P. and Stasik, M. (1997) "Design consideration for space and time distributed collaborative spatial decision making", *Journal of Geographic Information and Decision Analysis*, 1(1): 1–8. Available at: http://publish.uwo.ca/~jmalczew/gida.htm. Accessed on 4/25/00.

Lake, R.W. (ed.) (1987) *Resolving Locational Conflict*, New Brunswick, NJ, Rutgers Center for Urban Policy Research.

Lewis, F. (1994) *A Brief Introduction To Group Support Systems*, Bellingham, WA. MeetingWorks Associates.

NOAA (National Oceanic and Atmospheric Administration) (1993) "Technical notes from the Elliot Bay/Duwamish restoration program", April 14, 1993. *NOAA Restoration Program*, Sand Point Office, Seattle, WA.

Mosvick, R. and Nelson, R. (1987) *We've Got To Start Meeting Like This: A Guide To Successful Business Meeting Management*, Glenview, Ill., Scott, Foreman.

Nunamaker, J.D.A., Valacich, J., Vogel, D. and George, J. (1993) "Group support systems research: experience from the lab and the field", in L. Jessup and J. Valacich (eds) *Group Support Systems: New Perspectives*, New York, Macmillan Publishing Company.

Nyerges, T.L. (1995) "Cognitive task performance using a spatial decision support system for groups", in T. Nyerges, D.M. Mark, R. Laurini and M. Egenhofer (eds) *Cognitive Aspects of Human-Computer Interaction for Geographic Information Systems*, Proceedings of the NATO ARW, Mallorca, Spain, March 21–25, 1994, Dordrecht, Netherlands, Kluwer Academic Publishers: 311–23.

Orsborne, J., Moran, L., Musselwhite, E. and Zenger, J. (1990) *Self-Directed Work Teams: The New American Challenge,* Homewood, Ill., Business One Irwin.

Rittel, H.W.J. and Webber, M.M. (1973) "Dilemmas in a general theory of planning", *Policy Sciences*, 4: 155–69.

Sage, A. (1991) "An overview of group and organizational decision support systems", *IEEE Control Systems Magazine*, August 1991: 29–33.

Shiffer, M.J. (1992) "Towards a collaborative planning system", *Environment and Planning B*, 19(6): 709–22.

Shiffer, M. (1995) "Geographic interaction in the city planning context: beyond the multimedia prototype", in T. Nyerges, D.M. Mark, R. Laurini and M. Egenhofer (eds) *Cognitive Aspects of Human-Computer Interaction for Geographic Information Systems*, Proceedings of the NATO ARW, Mallorca, Spain, March 21–25, 1994, Dordrecht, Netherlands, Kluwer Academic Publishers: 295–310.

# 2 Macro-micro framework for participatory decision situations

**Abstract**

This chapter introduces the reader to a macro-micro approach to decision processes. It is a systematic yet flexible approach for characterizing complex geographic decision making, and one way of setting a foundation for understanding the complex character of decision support opportunities. We provide an example of a macro-micro decision strategy as a way of expressing the core issues in the macro-micro approach. Once the basic macro-approach has been presented, we elaborate on the micro aspect of understanding complex decision situations in terms of a revised version of Enhanced Adaptive Structuration Theory (EAST) — what we now call EAST2. EAST2 is composed of 25 aspects collected into eight constructs. Relationships between the eight constructs are described in terms of seven premises. We show how the premises can motivate research questions to focus empirical studies about participatory geographic information systems use. We have used EAST2 to guide us in our empirical research investigations involving GIS-supported collaborative decision making, as reported in Chapters 5–7.

A macro-micro approach to decision situations helps us understand complex geographic decision making and provides a way of characterizing decision support opportunities. In this chapter we provide an example of a macro-micro decision strategy as a way of expressing the core issues in the macro-micro approach. Once the basic approach has been presented, we elaborate on a way of understanding complex decision situations in terms of a revised version of Enhanced Adaptive Structuration Theory — what we now call EAST2. EAST2 is composed of twenty-five aspects collected into eight constructs. Each of these constructs and respective aspects is discussed below. Relationships between constructs and hence aspects are described in terms of seven premises as the foundation for motivating research questions about human-computer-human interaction during PGIS use. Research questions focus empirical studies about PGIS use.

## 2.1 Macro-micro approach to decision situations

People interacting within and among departments, organizations, agencies, governments, etc. encourage each other to address complex, geographic public-private, decision situations, as introduced in Chapter 1. How they go about it stems from what might have worked for them before, or not, and as communicated through people and reports. The way people bring together information (or do not as the case may be) might be "tried and true", or it might be "something novel". Such strategies might follow a well thought-out agenda or such processes might proceed ad hoc — even ad hoc approaches are implicit strategies. That process of what is, or is not working, what has, or has not worked, can be called a "macro strategy" in a decision situation. As the foundation for a conceptual framework, such a strategy might have been specified in a normative manner, i.e. what should happen. Perhaps the strategy developed through time in an incremental manner, tracked through decision chain analysis (Pressman and Wildavsky 1984), and now expressed as a non-redundant sequence of steps. The rationale behind the strategy might have been based on a variety of social values for process effectiveness to encourage one or more perspectives — consensual, empirical, political, and/or rational — to predominate (Reagan and Rohrbaugh 1990). Framing the many aspects of such decision situations by way of macro-micro strategies, as we highlight the use of participatory geographic information systems (PGIS), is the topic of this chapter.

After 30 years of work in GIS for landscape design and planning, Steinitz (1990) describes how a conceptual framework can be used to tie together six levels of inquiry; each level is associated with a type (phase) of modeling with GIS to form a comprehensive expression of a decision support strategy for landscape planning and design. Steinitz (1990) recognizes that the conceptual framework can tie together not only the modeling but also the theories that underlie the models, in that each model is "theory-driven". We use that same sense of a conceptual framework as a way of articulating a macro-micro strategy for a decision situation that could be composed of any number of macro-phases. We draw upon the work of several researchers who describe excellent examples of macro-level decision strategies for decision situations in landscape planning and design (Steinitz 1990), citizen participation in environmental decision making (Renn *et al.* 1993), several topics of environmental decision making (National Center for Environmental Decision-Making Research 1998), community-based redevelopment of brownfields (Electrical Power Research Institute 1998), and environmental and public health risk-informed decision making (Stern and Fineberg 1996).

In addition to the macro-level tasks of a decision situation, there are many "embedded task details" that appear to be characteristic of almost all decision situations we have come across. To help explain the relationships among details of decision situations we draw upon enhanced adaptive structuration theory (EAST) developed by Nyerges and Jankowski (1997) and its updated version, EAST2, as developed in section 2.3. We use EAST2 to characterize each macro-phase of a decision situation (i.e. from one macro-phase to another macro-phase), but with different instantiations of concepts for each phase. As such, we provide a characterization of decision situations using "micro strategy" details linked to the macro strategy; thus we articulate a macro-micro conceptual framework. This macro-micro approach with EAST2 further explains the social-behavioral implications of PGIS use in society. As such, the macro-micro approach further helps with "making a theoretical turn" in geographic information science (Pickles 1997), as a turn from focusing on just technology to focusing on technology in a social context.

While we use a macro-micro approach to examine decision situations, we want to note that we see a difference between a conceptual framework and a theory. The distinction is an important one in our exposition of participatory decision making. When the concepts and relationships in a conceptual framework are refined to a point whereby the relationships are referred to as premises (motivating research questions and propositions/hypotheses), then the conceptual framework would be called a theory. Consequently, the difference between a conceptual framework and a theory is that relationships play a descriptive role in conceptual frameworks, but an explanatory role in a theory.

Some theories might involve only one premise, while other more comprehensive theories might involve multiple premises. For the macro-phases of a decision situation, which seem to be dependent on "particular agendas" for what works, we do not (yet) claim to have elucidated a theory for the macro level of decision making, i.e. a theory for explaining (project) macro-level decision phases. The dynamics of participatory settings are not as clearly understood for explanations to be put forward, but EAST2 is a robust conceptual framework for that level. In this regard we are simply being conservative with our claims. Nonetheless, because we use EAST2 at the micro level to understand "task oriented" decision making based on multiple premises, what we have created is a very powerful macro-micro framework. The entire framework we elucidate is thus termed a "conceptual framework for macro-micro decision situations". This framework has provided us with significant insight into the regularity and idiosyncrasies of a number of complex, geographic decision situations. Certainly, it is not the only framework that could be used to explain the complexity of participatory decision situations, but we will show it is one of the most comprehensive, and useful in many regards as we use it often in the remaining chapters of this book.

The EAST2 framework provides us with an understanding that there are at least eight major perspectives on how to interpret the character of a decision situation. There is one for each of the EAST2 constructs which are:

- social-institutional influences that motivate people and/or organizations to address a concern;
- group participant influence within a group of people coming together in a situation;
- participatory GIS influence as tools that could be used to address a concern;
- actual appropriation of tools within the task activity;
- group process in terms of task and conflict management;
- emergence of information structuring during group processes;
- task outcomes for a decision(s);
- social outcomes from addressing a task.

What is interesting about this array of constructs is that they represent at least five "different perspectives" that are commonly used to describe decision situations. The social-institutional construct is responsible for a "functional" perspective on decision making, i.e. types of decisions characterized as operational (shorter-term), tactical (medium-term), and strategic (longer-term) decision situation horizons. The group participant construct is associated with an "organizational" perspective about the group being convened, i.e. "who is coming to the table" to take part in the decision situation. The PGIS information construct is the core of the "tool" perspective that provides a sense of what information capabilities are available to provide information insight into the decision situation. As the above three perspectives (constructs) combine together to influence the process used to address decision situations, we can say a fourth perspective is a "participatory" decision perspective, i.e. how social-institutional influences, group participant influence, and information structure influences combine from one task to the next in a flow of decision making. A fifth perspective deals with the outcomes of a decision situation, i.e. both decision outcomes as well as social outcomes. The success or failure of such outcomes is very likely to help/hinder people's views about carrying through, aborting, or ever again entering into similar kinds of decision situations — the crux of social reconstruction.

We turn next to present an overview in section 2.2 of an example macro-micro decision strategy. The example sets the stage for the discussion about the constructs and premises of EAST2 in section 2.3 that provide explanatory insight for the perspectives mentioned above.

## 2.2 Example of a macro-micro decision strategy

The use of software for supporting any decision situation requires some amount of regularity. At some level of decision support system design there is always a tension between regularity as constraint versus regularity as enabling stability. At the macro level of decision support we assume that some kind of decision process agenda can be set by a group to address concerns. By decision process agenda we mean a normative, but flexible, structure; here normative implies "what might be likely to happen". We call such an agenda a "decision strategy" for a particular project, as set by the authority or group who convenes to address a decision problem. The benefit of a structured understanding of the decision process is that it lends itself to computer support, hence the regularity. Whether a group follows the agenda (i.e. strategy) is another issue, which is why we have chosen this macro-micro approach to characterizing decision situations. Nonetheless, the main challenge for any agenda structuring approach to spatial decision making process is to make it flexible enough so that a variety of spatial decision problems can be accommodated.

In essence, the macro-micro framework stems from recognizing that the external and internal influences on decision situations are likely to encourage a decision process that fits the nature of the decision problem at hand, rather than imposing a "one-process fits all problems" kind of process. The macro aspect of this framework recognizes that decision situations are interpreted by groups in different ways, hence they (groups) are likely to set an agenda based on what they feel is appropriate to the problem. As such, our macro-micro framework synthesizes a wide array of frameworks about decision making, including frameworks from literature concerning: managerial (Simon 1977, 1979), organizational computing (Bhargarva, Krishnan and Whinston 1994), public participation (Renn *et al.* 1993), collaborative GIS-support (Nyerges and Jankowski 1997), landscape planning (Steinitz 1990), environmental (National Center for Environmental Decision-Making Research 1998), community-based (Electrical Power Research Institute 1998), and environmental risk-informed (Stern and Fineberg 1996) aspects of decision making. In the case of managerial decision making, Simon (1977, 1979) recognizes that the four steps of intelligence, design, choice, and review are essential tasks of individual decision making in an organizational context. Renn *et al.* (1993) have used a three-step process — criteria development, options generation, and options evaluation — in public participatory decision making in both the USA and Germany to help recommend environmental policy. Steinitz (1990) sees six steps in modeling for landscape planning which include representation models, process models, evaluation models, change models, impact models, and decision models. The National Center for Environmental Decision-Making Research (1998) describes a sequence of tools for data gathering and analysis that could include more or fewer steps depending on the situation at hand. The Electrical Power Research Institute (1998) recommends using their SmartPlaces Series E GIS (oriented to community development decision making) as part of a ten-step process. The National Research Council effort concerning analytic-deliberative decision making about risk-oriented hazards, suggested that multiple groups be involved in such processes, and that the processes should be "organic" in character, i.e. determined by the groups being convened

(Stern and Fineberg 1996). The point is that each of the above strategies consists of a variable number of steps that are appropriate for a variety of decision situations.

One can synthesize all of the above successful frameworks into a single macro-step framework, which has often been done by various researchers. But a new synthesis would suggest a question: why bother? Each framework, offering a different successful strategy, seems to do what it does best in the particular context for which it was developed, by being articulated in terms of the number of phases that work! Thus, the macro part of the framework developed here adopts whatever set of steps would seem to work for the particular situation at hand. All of the above decision processes as given sequences of steps have been tested in practice. Thus, "not just any" macro-step process should be adopted. When a given sequence of steps seems to work for a particular (decision) problem situation, one can call that sequence of steps a "decision strategy", in the sense of an agenda that has "seemed" to work. However, this macro-micro framework would not be anything new if we stopped here and left the issue by simply recommending a "use what works" strategy.

There is something systematic about each of the steps in the above decision frameworks. Bhargarva, Krishnan and Whinston (1994) were the first to articulate (at least to the authors' knowledge) the idea that Simon's (1977) steps of intelligence, design and choice, each had within them an iterative process of intelligence, design and choice. That is, when undertaking an intelligence process for decision making, it is natural that people pursue intelligence (i.e. gathering information), design (i.e. organizing information), and choice (i.e. selecting information). The same would occur for the subsequent steps of design and choice, i.e. each has within it, intelligence, design and choice. As such, complex decision processes are recursive in nature at both macro and micro levels. A point that was missed by Bhargarva, Krishnan and Whinston (1994) is that various organizations, and even individuals and groups within those organizations, might actually use a different macro-strategy during inter-organizational work (Nyerges and Jankowski 1997); that is, work between an individual, group and/or organization, and other external individuals or groups or organizations is different, depending on the situation as by reference to the no less than six practical working strategies cited above. Such inter-organizational work, sometimes known as coalitions, alliances, work-groups etc, have always occurred, but are on the increase due to the recognition that complex problems often require different collaborative approaches to decision making (Gray 1989). Although different macro-strategies are likely to be appropriate for different contexts, nonetheless, a micro-strategy is at play for each macro-step associated with a work task. That micro-step strategy, encouraged by fundamentals of human information processing, includes gathering, organizing, selecting, and reviewing information (Simon 1977). Furthermore, we would be remiss if we did not mention Dewey's (1933) work on reflective thinking that is relevant to every macro-micro step: a reflective thinking process promotes "active learning" along the way in complex decision situations. After all, there is usually no single person who knows all about the complex situation, otherwise we would simply task that person to work out a solution.

The macro-micro perspective allows us to appreciate that every macro-phase in a macro strategy can have a different set of information needs, based on the collective needs of the micro-step activities. Consequently, a macro-micro decision strategy motivates (in large part) the requirements for decision support tools (discussed in Chapter 3). Such information needs (as well as organizational, social, individual and other needs as they arise) and the associated requirements for decision support tool, can only be addressed by a good understanding of the decision situation at the time and place (context) within which it occurs. This then has been the major stumbling block in group-based decision support, i.e. a flexible but thorough framework for unpacking the complexity of needs from a macro-micro perspective has not been proposed before. We do that here by way of example.

We synthesize the Renn *et al.* (1993) three-step public-participation decision process with the Simon (1977) three-step process for the macro-level of a macro-micro decision strategy (see matrix of Table 2.1). In this example we have a strategy consisting of (1) intelligence about values, objectives, and criteria, to form a value tree (2), design of a feasible option list, and (3) choice about recommendations as listed across the columns of the matrix. The combination of the Renn *et al.* (1993) and Simon (1977) macro-phasing shows the similarity in the two decision processes, without doing disservice to either. It is important to point out that any of the other six macro-strategies could have been used for our explanation. If we had adopted the ten-phase SmartPlaces macro decision strategy (Electrical Power Research Institute 1998) then there would be ten columns across the top, one for each major phase in their process. If we had adopted the Steinitz strategy (1990) we would have used six phases, one for each of the modeling steps, across the columns. We chose the Simon (1977) and Renn *et al.* (1993) public participation strategies because of the fewer number phases that conveniently allow us to present our macro-micro approach and because this process is also generally applicable to the decision situations on which we report in Chapters 5–7. Of course, there are similarities and differences overall, but the basic approach is similar given that the three decision situations in Chapters 5–7 can be called "site selection situations". As such, our discussion directly to follow draws in general from the habitat site selection process presented in Chapter 7

Regardless of the number of macro-phases, the micro aspect depicts a regularity to which we referred earlier. At a micro-activity level, i.e. within

*Table 2.1* An example macro-micro, participatory decision strategy

| | *Macro-phases in a decision strategy* | | |
|---|---|---|---|
| *Micro-activities in a decision strategy* | *1. **Intelligence** about values, objectives and criteria* | *2. **Design** of a set of feasible options* | *3. **Choice** about recommendations* |
| **A. Gather ...** | issues to develop and refine **value trees** as a basis for objectives | **primary criteria** as a basis for option generation | **values, criteria, and option list scenarios** for an evaluation |
| **B. Organize ...** | **objectives** as a basis for criteria and constraints | and apply approach(es) for **option generation** | approaches to **priority and sensitivity analyses** |
| **C. Select ...** | **criteria** to be used in analysis as a basis for generating options | the **feasible option list** | **recommendation** as a prioritized list of options |
| **D. Review ...** | criteria, **resources, constraints,** and **standards** | **option set(s)** in line with resources, constraints and standards | **recommendation(s)** in line with original **value(s), goal(s)** and **objectives** |

each step as described above, there are at least four decision activities: gather, organize, select and review (Simon 1977, Bhargarva, Krishnan and Whinston 1994 only mention three), with reflection for every activity (Dewey 1933). The micro strategy for any macro task thus becomes:

A   gather information;
B   organize that information;
C   select from that information; and
D   review what information is really needed to move on to the next phase.

When reading the matrix first note the column headings, then read the verbs in the very left column, e.g. for the Intelligence phase, the first activity-phase is A. Gather issues to develop and refine value trees, B. Organize objectives, C. Select criteria, D. Review resources, constraints and standards; then for the Design phase, A. Gather primary criteria, and so on.

Although the micro-activity strategy repeats itself for each phase, giving a sense of stability, it is important to remember that each phase has a different task associated with it, so the process for each macro-phase is actually very different. In fact, in the Renn *et al.* three-phase process, different groups were convened to undertake each macro-phase as they were in the habitat site selection problem. In the first phase stakeholders were consulted for values elicitation through interviews and meetings to arrive at a set of basic objectives that could lead to criteria. In the second phase technical specialists were consulted for options generation, and a feasible set of land parcels along the Duwamish Waterway were identified. In the third phase, technical specialists ranked sites and then forwarded them to the decision board for finalization. In a similar vein, the reader will see in Chapters 5, 6, and 7, that although the same general process applies, the macro-phases for rural primary health care funding allocation, transportation improvement project site selection, and habitat development site selection, respectively, are indeed rather different among phases. Being able to accommodate different aspects of tasks and perspectives from groups, as well as many other aspects in EAST2 from phase to phase, is part of the flexibility of this approach, as described in detail in section 2.3.

The four micro activities together with three macro phases of the decision process constitute twelve "phase-activities" of this particular version of the macro-micro framework — remembering that it can be very different from one practical decision situation to another. The significance of the labeling "phase-activity" is that a phase refers to the issue of what is expected as an outcome in the overall strategy, while an activity is an action that takes place to facilitate creation of the outcome. Although the term "step" can be used, as has been convenient in much literature, phased-activities provide a nuance that is critical in getting work done. In general, groups are likely to move through these activity-phases starting from the upper left-hand cell (1A) through to (1D), moving on to (2A–2D), and finally through (3A–3D). The intelligence phase generally consists of problem discussion/definition that leads to a set of measurable criteria used to evaluate various options as solutions to the problem. As part of this definition in the intelligence phase, a person, group, or community might articulate values, goals, objectives, and criteria about the character of a decision problem (Keeney 1992). Personal, organizational and/or community values could be articulated as "background" reasons for why a goal is important from various perspectives. A goal essentially describes why a group feels a problem situation is important to deal with at all. Articulating values and goals leads to a set of objectives for defining the important characteristics that an option should have. One or more criteria stem from each of the objectives, each of the criteria being a measurable characteristic on which the option can be evaluated. Von Winterfeldt (1987) used value trees to compare and contrast the varying interests of stakeholder groups in a decision process concerning oil leases off the coast of California.

In the design phase, a problem is structured by identifying feasible options, part of which involves finding a suitable way to assess feasibility and to identify a technique that supports option evaluation. Options generation is a fundamental analysis problem in GIS. It commonly relies upon establishing one or more primary attribute(s) as a practical way to control the identification and subsequent differentiation of options. The challenge lies in understanding how "entity categories" are created for GIS database designs (Nyerges 1991). For example, in case of habitat projects, a land parcel boundary is used as a way of differentiating land by ownership (i.e. legal custodial responsibility). Thus, the "polygon" character of a land parcel allows us to differentiate one entity of land from another as a basis of searching a geographic space for feasible options. Even if we used parcel (polygon) centroids as the "place holder" for parcels, it is the boundary that distinguishes one potential option from another. If in the case of rural parcels that are usually rather large, where boundaries are not of a spatial resolution to make sense for resolving habitat, then we might adopt a "raster cell" unit as the basis of enumerating options, so that each area on the ground contained within a raster cell forms the basis of a potential option. The controlling character of an option is dependent on the problem under consideration, but in geographic problems it is likely to be one or more of a combination of points, lines, polygons, or raster cells. However, it is equally important that the next task is then to add secondary attributes (criteria) as part of the information to be considered, so that the option set (collection) can be expanded or contracted, depending on what is practical, given the phenomenon of interest. What makes an option feasible is a matter of setting a minimum and/or maximum threshold for the primary or secondary attribute, e.g. a practical range of size or condition of ownership, so as to be able to include it into a set of parcels for further consideration. The feasible set of options once enumerated could be thought of as a "data space for what is possible". In GIS terms, Boolean operations in a query language can be used to "filter" the total option set based on inclusionary (to include within the option set) or exclusionary (to exclude from the option set) criteria.

The choice phase involves an evaluation of the options in terms of the criteria identified in Phase 1. What this really means is that the options are evaluated in terms of the criteria that stem from the objectives that stem from the values identified in Phase 1. If a goal, or values or objectives were not established, then a person, group or community will end up having this conversation at this point, because there will be no other practical means of comparing options. It is possible that many scenarios might be developed. The different scenarios can be based on the various option sets. The options sets can be simply clusters of options deemed important for consideration based on different thresholds for criteria or geographic location. The option evaluation commonly occurs in the form of integrating criterion data values, measuring the observed or predicted outcomes of each feasible option on every criterion, with preferences (weights on criteria) set by the decision participants. As a follow-up to this integration function, a sensitivity analysis can be performed to see how the options might be evaluated differently if the weights are changed. Options are often ordered from most to least desirable, after which options are recommended. Sometimes, however, we are interested in finding an option or set of options that meets a minimum threshold. Thus, ordering from best to least desirable would not be necessary. Reviewing the process gives the participants a chance to see how the phase-activities play out overall. Chapter 3 deals with more details of the methods that can be used for criteria identification, option generation and option evaluation.

The macro-micro decision strategy presented above amounts to a normative description of an expected decision process, or at least a good idea of one from a rational perspective. It exists essentially as an efficient way to organize people energy. However, when performing social-behavioral studies of decision processes, that normative strategy can form the basis of an empirical coding system of what actually transpires in a decision process, since we would not expect most groups to follow a normative process. When describing what actually transpires from macro-phase to macro-phase researchers have used decision chains, consisting of steps for which there is a decision point and one or more clearance points to move to the next step (Pressman and Wildavsky 1984, Drew *et al.* 2000). Along this train of thinking, an interesting research question that has continually confronted us is: if groups need to visit every phase-activity in order to get work accomplished, why do some groups move through the process in one way, whereas others move through it in another? In many decision situations some of the phase-activities might not necessarily be addressed explicitly, i.e. some of the activities are pursued without consciously documenting every activity, whereas in other instances, full documentation might be provided in memos, notes or reports. So another interesting empirical research question is: if some activities are not addressed explicitly, is this a major reason why groups must back-track and revisit activities when they find a lack of information available for their use? To address such questions, we use macro-micro strategies as described here as the basis of "decision function coding schemes" to unpack the character of decision processes in empirical research. We will discuss this further in Chapter 7; but see also Nyerges *et al.* (1998) and Drew *et al.* (2000) for more information about empirical coding of decision processes.

In the next section we present EAST2 as the theory providing the context for the PGIS methods presented in Chapter 3, the backdrop for the empirical research strategies discussed in Chapter 4, and the foundation of the empirical research reported in Chapters 5–7.

## 2.3  Conceptual foundations — Enhanced Adaptive Structuration Theory 2

We described the commonality between a conceptual framework and a theory in section 2.1. Both involve a set of concepts and a set of relationships among those concepts to organize our thoughts about the world. The difference between them, however, is that the relationships in a theory take on an "explanatory" level of meaning, rather than just description. However, such relationships as possible explanations need to be examined (tested) to develop empirical findings to see whether the theory provides a "useful organization" of ideas to enhance our understanding about the world.

From a theoretical perspective, choice of a theory (or building a theory) for articulating what to expect during human-computer-human interaction provides a way of "systematically" interpreting how people make use of GIS in a problem context. In our work we started with Adaptive Structuration Theory (AST) to provide a framework for studying group decision making in an organizational context (DeSanctis and Poole 1994). AST was developed to explain human-computer-human interaction that incorporates advanced information technology, specifically group decision support systems, in a face-to-face computer network setting. AST consists of a set of eight constructs, as the basic *elements* of the theory, that outline significant issues for characterizing group decision making, and a set of seven premises that describe the *relations* between the eight constructs. Part-way through the research Nyerges and Jankowski (1997) found it necessary to develop Enhanced AST (EAST) to frame systematic examinations of complex, inter-organizational participatory processes that make use of PGIS technology (Figure 2.1). EAST treats a total of 21 aspects (indicated by "*" in Table 2.2). Those aspects were collected from among AST (DeSanctis and Poole 1994), several frameworks about GIS use (Calkins and Obermeyer 1991, Campbell and Masser 1995, Dickinson 1990, Obermeyer and Pinto 1994, Pinto and Azad 1994, van der Schans

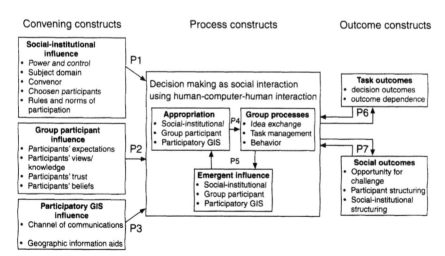

*Figure 2.1* Enhanced Adaptive Structuration Theory 2 (EAST2) frames convening, process, and outcome constructs plus the respective premises to provide a conceptual map for understanding a group decision support situation

*Table 2.2* Twenty-five aspects of EAST2\*: A theory of GIS-supported
participatory decision making

---

**Constructs and aspects about convening a participatory situation**

---

*Construct 1: social-institutional influence*

\*1)  **power and control** (Gray 1989, Habermas 1984, Kunreuther *et al.* 1983,
      Kunreuther, Slovic and MacGregor 1996, Ren, Webler and Wiedemann 1995,
      Susskind and Field 1996)

\*2)  **subject domain as task purpose, content, and structure** (Beck 1992, Coenen,
      Huitema and O'Toole 1998, Couclelis and Monmonier 1995, DeSanctis and
      Poole 1994, Functowicz and Ravetz 1985, Healey 1995, Kunreuther *et al.*
      1983, Lake 1987, McGrath 1984, Popper 1981, Raiffa 1982, Renn, Webler and
      Wiedemann 1995, Rittel and Webber 1973, Susskind and Cruikshank 1987,
      Wood and Gray 1991)

\*3)  **persons, groups, and/or organizations as convenor of participants** (Coenen,
      Huitema and O'Toole 1998, Cowie and O'Toole 1998, Habermas 1984, Renn,
      Webler and Wiedemann 1995, Stern and Fineberg 1996, Susskind and Field
      1996, Susskind, Levy and Thomas-Larmer 2000, Wood and Gray 1991)

\*4)  **choosing the number, type and diversity of participants** (Campbell 1996,
      Campbell and Masser 1995, Cowie and O'Toole 1998, DeSanctis and Poole
      1994, Gray and Wood 1991, Habermas 1984, Harris and Weiner 1998, Healey
      1995, Kunreuther *et al.* 1983, Ostrom 1992, Pasquero 1991, Poole 1985, Raiffa
      1982, Reagan and Rohrbaugh 1990, Renn, Webler and Wiedemann 1995,
      Sackmann 1991, Stern and Fineberg 1996, Roberts and Bradley 1991, Wood
      and Gray 1991)

\*5)  **rules and norms as social structures among participants** (Campbell 1996,
      DeSanctis and Poole 1994, Habermas 1984, Healey 1995, Kunreuther *et al.*
      1983, Pinto and Azad 1994, Raiffa 1982, Renn, Webler and Wiedemann 1995,
      Wood and Gray 1991)

*Construct 2: group participant influence*

\*6)  **participants' expectations based on values, goals, issues, values, beliefs, and
      fairness** (Couclelis and Monmonier 1995, Davy 1996, DeSanctis and Poole
      1994, Habermas 1984, Kunreuther *et al.* 1983, Kunreuther, Slovic and
      MacGregor 1996, Leitner *et al.* 1998, Linnerooth-Bayer and Löfstedt 1996,
      Lober 1995, McGrath 1984, 1991, Obermeyer and Pinto 1994, Ostrom 1992,
      Raiffa 1982, Reagan and Rohrbaugh 1990, Susskind and Cruikshank 1987,
      Susskind and Field 1996, Wood and Gray 1991)

\*7)  **participants' views/knowledge of the subject domain and each other**
      (DeSanctis and Poole 1994, Habermas 1984, Kahneman 1992, Keeney 1992,
      Raiffa 1982, Susskind and Field 1996, Wood and Gray 1991)

\*8)  **participants' trust in the process** (Coenen, Huitema and O'Toole 1998, Duffy,
      Roseland and Gunton 1996, Kunreuther, Slovic and MacGregor 1996,
      Susskind and Field 1996)

\*9)  **participants' beliefs and feelings toward information technology** (DeSanctis
      1993, Shiffer 1998, Turk 1993)

*Construct 3: participatory GIS influence*

\*10)  **place, time, and channel of communications** (Habermas 1984, Healey 1995,
       Jarke 1986, Leitner *et al.* 1998, Nyerges 1995, Renn, Webler and Wiedemann
       1995, Wood and Gray 1991)

*Table 2.2* Continued

---

\*11) **availability of social-technical structures as information aids** (Armstrong 1993, Calkins and Obermeyer 1991, Campbell and Masser 1995, Densham 1991, DeSanctis and Poole 1994, Dickinson 1990, Faber, Wallace and Cuthbertson 1995, Jankowski and Stasik 1997, Jarvenpaa 1989, Jarvenpaa and Dickson 1988, Kunreuther *et al.* 1983, Harris and Weiner 1998, Healey 1995, McGrath 1990, Moore 1997, Obermeyer 1998, Obermeyer and Pinto 1994, Pinto and Azad 1994, Shiffer 1998, Susskind and Cruikshank 1987, van der Schans 1990)

---

**Constructs and aspects about participatory process as social interaction**

*Construct 4: appropriation*

12) **appropriation of social-institutional influence** (DeSanctis and Poole 1994, Gray 1989, Habermas 1984, Healey 1995, Kunreuther *et al.* 1983, McGrath 1984, Pinto and Azad 1994, Raiffa 1982, Susskind and Field 1996, Wood and Gray 1991)

13) **appropriation of group participant influence** (Couclelis and Monmonier 1995, DeSanctis and Poole 1994, Habermas 1984, Kunreuther *et al.* 1983, Kunreuther, Slovic and MacGregor 1996, Leitner *et al.* 1998, McGrath 1984, 1991, Obermeyer and Pinto 1994, Raiffa 1982, Roberts and Bradley 1991, Susskind and Field 1996, Wood and Gray 1991)

\*14) **appropriation of participatory GIS influence** (Calkins and Obermeyer 1991, Campbell and Masser 1995, DeSanctis and Poole 1994, Dickinson 1990, Kunreuther *et al.* 1983, Healey 1995, McGrath 1990, Obermeyer and Pinto 1994, Pinto and Azad 1994, Susskind and Cruikshank 1987, van der Schans 1990)

*Construct 5: group process*

\*15) **idea exchange as social interaction** (Campbell and Masser 1995, DeSanctis and Poole 1994, Habermas 1984, Healey 1995, Kunreuther *et al.* 1983, McCartt and Rohrbaugh 1995, McGrath 1990, 1991, Nyerges *et al.* 1998, Obermeyer 1998, Raiffa 1982, Reagan and Rohrbaugh 1990, Renn, Webler and Wiedemann 1995, Roberts and Bradley 1991, Stern and Fineberg 1996, Susskind and Cruikshank 1987, Susskind and Field 1996, Wood and Gray 1991)

\*16) **participatory task flow management** (Bhargarva, Krishnan and Whinston 1994, Coenen, Huitema and O'Toole 1998, DeSanctis and Poole 1994, Gray 1989, Healey 1995, Kunreuther *et al.* 1983, McCartt and Rohrbaugh 1995, McGrath 1990, 1991, Nyerges *et al.* 1998, Poole and Roth 1989, Raiffa 1982, Reagan and Rohrbaugh 1990, Renn *et al.* 1993, Renn, Webler and Wiedemann 1995, Stern and Fineberg 1996, Susskind and Cruikshank 1987)

\*17) **behavior of participants toward each other** (DeSanctis and Poole 1994, Habermas 1984, Harris and Weiner 1998, Healey 1995, Kunreuther *et al.* 1983, Nyerges and Jankowski 2001, Raiffa 1982, Susskind and Cruikshank 1987, Susskind and Field 1996, Wood and Gray 1991)

*Construct 6: emergent influence*

\*18) **emergence of PGIS influence** (Calkins and Obermeyer 1991, DeSanctis and Poole 1994, Dickinson 1990, Healey 1995, Kunreuther *et al.* 1983, McGrath 1990, Obermeyer 1998, Obermeyer and Pinto 1994, Shiffer 1998, van der Schans 1990)

*Table 2.2* Continued

---

\*19) **emergence of group participant influence** (DeSanctis and Poole 1994, Healey 1995, Renn, Webler and Wiedemann 1995, Roberts and Bradley 1991, Wood and Gray 1991)

20) **emergence of social-institutional influence** (DeSanctis and Poole 1994, Harris and Weiner 1998, Healey 1995, Kunreuther *et al.* 1983, McGrath 1984, Raiffa 1982, Roberts and Bradley 1991, Stern and Fineberg 1996, Wood and Gray 1991)

**Constructs and aspects about participatory outcomes**

---

*Construct 7: task outcomes*

\*21) **character of decision outcomes** (Chun and Park 1998, Coenen, Huitema and O'Toole 1998, Cowie and O'Toole 1998, DeSanctis and Poole 1994, Dickinson 1990, Duffy, Roseland and Gunton 1996, Healey 1995, Jessup and Valacich 1993, McCartt and Rohrbaugh 1995, McGrath 1984, 1990, 1991, Obermeyer and Pinto 1994, Ostrom 1992, Ostrom, Gardner and Walker 1994, Pestman 1998, Pressman and Wildavsky 1984, Reagan and Rohrbaugh 1990, Renn, Webler and Wiedemann 1995, Webler 1995)

\*22) **decision outcome and participant structuring dependence** (Burns and Ueberhorst 1988, Coenen, Huitema and O'Toole 1998, McGrath 1990, 1991, Raiffa 1982, Susskind and Cruikshank 1987, Susskind and Field 1996, Wood and Gray 1991)

*Construct 8: social outcomes*

\*23) **opportunity for challenge of the outcome** (Habermas 1984, Healey 1995, Pressman and Wildavsky 1984, Renn, Webler and Wiedemann 1995, Stern and Fineberg 1996, Susskind and Field 1996)

\*24) **reproduction and temporality of group participant influence** (Coenen, Huitema and O'Toole 1998, DeSanctis and Poole 1994, Habermas 1984, McGrath 1990, 1991, Pestman 1998, Pretty 1998, Renn, Webler and Wiedemann 1995, Susskind and Cruikshank 1987, Susskind and Field 1996, Wood and Gray 1991)

25) **reproduction and temporality of social-institutional influence** (DeSanctis and Poole 1994, Habermas 1984, McGrath 1990, 1991, Ostrom 1992, Ostrom, Gardner and Walker 1994, Roberts and Bradley 1991, Susskind and Cruikshank 1987, Susskind and Field 1996, Wood and Gray 1991)

---

\*Original aspects of EAST in Table 1 of Nyerges and Jankowski (1997).

1990), collaboration theory (Wood and Gray 1991), participatory negotiation theory (Susskind and Field 1996), political negotiation theory (Kunreuther *et al.* 1983), and communicative action theory (Habermas 1984, Healey 1995). Refining our theoretical turn to a social context for PGIS technology use in society (Pickles 1997), EAST2 further explicates the character of inter-organizational decision processes for public–private contexts by treating 25 aspects of group decision making (see Table 2.2) — the original 21 of EAST, plus four more. Here we have re-synthesized the 21 aspects and introduced four additional aspects from literature not considered in our earlier work, including concepts from rational choice theory (Ostrom 1992, Ostrom, Gardner and Walker 1994), competing values theory (McCartt and Rohrbaugh 1995, Reagan and Rohrbaugh 1990), risk-based analytic-deliberative decision processes which reference GIS use (Stern and Fineberg 1996), citizen participation processes (Renn, Webler and Wiedemann 1995), as well as literatures about the use of public participation GIS (Harris and Wenier 1998, Leitner *et al.* 1998, Obermeyer 1998, Shiffer 1998).

EAST2 still retains the same seven premises of AST (the P's in Figure 2.1 described in section 2.3.2), since we have not added any new constructs. As in AST, each of the seven premises in EAST2 represent a fundamental statement about how constructs (and thus aspects) influence each other, hence provide expected "explanatory power" in the theory. However, we have broadened our treatment of the premises to include the eight additional aspects. In addition, we have reordered the appearance of constructs in the framework so that advanced information technology is now only one of the eight constructs, providing a different balance of treatment for constructs. In any decision situation, consideration of social-institutional and participation concerns by people generally precede consideration of technology — hence the reordering. Due to the introduction of eight more aspects, several other relationships between pairs of aspects are in need of empirical exploration when it comes to complex, inter-organizational, geographical, problem solving and decision making in anywhere, anytime meetings across multiple-task decision situations. All of these constructs and premises taken together constitute the structuration context of EAST2.

Structuration is the *embedding context* for (E)AST2, but from a post-structurationist perspective. Structuration is a process at play within and among existing mandates, people, and social-technical influences that organizes (inter-organizational) activity in various ways (Giddens 1984, Orlikowski 1992). The fundamental motivation for participants in the structuration process is their intentional attempts to (mis)understand each other through communicative action (Habermas 1984). The "intentional agency" of the participants of a group can influence changes in structural relationships at any time as appropriate to the situation. Agents of change as individuals, groups, organizations, or coalitions direct human effort by various means toward various ends for various reasons. Such agents are both assisted and/or hindered by various "structured" circumstances in life. Structures are then the relationships within and among participant groups, society, technology, and the social/physical environment. We assume that neither the technological nor social character of a situation predominates a priori in the structuring relationship, i.e. social-technical artifacts and the people making choices can encourage change in structure. By discipline the structural relationships could be characterized as social, economic, political, physical, etc. However, disciplines, particularly academic disciplines, are simply convenient lenses that people use to establish/interpret/redirect relationships inherent in laws, bureaucracies (as entrenched information flows), norms for social interaction, convenient ways of presenting information using various languages (graphic, verbal, written), emotional attachments, and natural environmental processes.

As a means of further elucidating the character of the structural influences in EAST2, and at the same time improving the readability of Figure 2.1, we have adopted Gray and Wood's (1991) three category scheme of organizing convening, process, and outcome elements of a collaboration. To help the reader move past what might seem a complex portrayal of aspects, we encourage him or her to view the depiction of EAST2 in Figure 2.1 as a macro-phase "task description" (with potential explanatory power) in line with the macro-micro strategy description in section 2.2 (Nyerges 1993). The diagram can be used to develop a comprehensive task description for each macro-phase task (e.g. "1. Intelligence" in Table 2.1) by describing the 25 aspects associated with that macro-phase task in whatever level detail is appropriate for the fundamental understanding of the situation. Essentially, each aspect becomes a question to pose about the nature of the situation. From macro-phase to macro-phase, any one, several or all of the 25 aspects might stay the same or change. The change of aspect(s) from one macro-phase to the next sheds light on macro-micro differences of the situation. Consequently, the "macro-micro" conceptual framework that embeds EAST2 is a powerful, yet flexible, organizing mechanism for framing research activity about "realistic" complex, inter-organizational decision situations. To further detail the descriptive power of EAST2 in the next section we describe each of the constructs and respective aspects. The section after that takes up the explanatory power by presenting the premises that relate the constructs (aspects) to each other.

### 2.3.1  Constructs of EAST2

We present each of the eight constructs and their respective aspects ordered by convening, process, and outcome categories. The order of discussion implies a loose dependence of conceptual issues. When EAST2 is used to frame empirical research studies, some aspects by way of the premises of concern and hence research questions posed are likely to be more appropriate than others for any particular study. How those aspects are actually made operational in terms of variables largely depends on social-behavioral data strategy as part of an empirical research design discussed in Chapter 4.

#### 2.3.1.1  Constructs and aspects about convening a participatory situation

Three constructs consisting of eleven aspects characterize the convening influence about a decision situation when information technology is involved.

*Construct 1: social-institutional influence*    Social-institutional influence is usually based in law, mandate, policy, social norm, or natural events, i.e. influences commonly outside the control of any single individual. One aspect of social-institutional influence is *power and control* that refers to the entitlements that are granted by formal or informal mandate, e.g. laws and regulations or special interest group awareness (Gray 1989, Kunreuther *et al.* 1983, Kunreuther, Slovic and MacGregor 1996, Susskind and Field 1996). This is an important topic from the research agenda of PGIS and society (Pickles 1995, Sheppard 1995), as well as participatory decision making (Kunreuther *et al.* 1983, Renn, Webler, and Wiedemann 1995). Whether specific participant groups intend to exercise their power is a concern to all participant groups (Susskind and Field 1996).

A second aspect of social-institutional influence is *subject domain as task purpose, content, and structure.* Wood and Gray (1991) point out that sharing an interest in "subject domain" is what principally brings people together into collaborative alliances to undertake inter-organizational group work. Sharing a subject domain is important for establishing a focus as to what constitutes a decision situation worth addressing, hence for why people come together to make decisions at all (DeSanctis and Poole 1994, Healey 1995, Kunreuther *et al.* 1983, McGrath 1984, Raiffa 1982, Wood and Gray 1991). Only since the mid-1980s have researchers been engaged with problems that motivate group-based software design of any nature. Focusing on small groups, McGrath (1984) carried out a comprehensive review of group activity literature to synthesize a task circumplex composed of eight types of tasks: generating ideas, generating plans, problem solving with correct answers, deciding issues with preference, resolving conflicts of viewpoint, resolving conflicts of interest, resolving conflicts of power, and executing performance tasks. Each of them differs in terms of content, goal and complexity in specific decision making situations.

In regards to task purpose, content, and complexity, it is now widely recognized that many spatial problems called "wicked" and "ill-structured" (Rittel and Webber 1973) contain intangibles that cannot be easily quantified, their structure is only partially known or burdened by uncertainties, and potential solutions often become NIMBY controversies (Couclelis and Monmonier 1995, Lake 1987, Popper 1981, Susskind and Cruikshank 1987). These problems require the participation of people representing diverse areas of competence, political agendas, and social values. Also, from an organizational standpoint, the specialized division of knowledge and skills in many communities means that complex decisions are formed through consideration of multiple input by people tasked with various activities and responsibilities in different locations. As a consequence, diverse groups often must generate solutions to pervasive spatial problems ranging across multiple administrative scales and geographic scales (Coenen, Huitema and O'Toole 1998).

Wicked, public–private problems are characteristic of at least three kinds of uncertainty which adds to the complexity (Coenen, Huitema and O'Toole 1998). One type of uncertainty relates to knowledge about the natural environment, i.e. what we do not know about natural processes and the influences that humans might have in such processes. A second type of uncertainty is about the intentions in related fields of choice from a technical perspective, i.e. whether one solution is technically better than another. A third type of uncertainty is about values, i.e. which valued concerns in society should be pursued, often with full awareness that we cannot pursue all values of concern. Beck (1992) observes that major social conflict in western societies has become centered around the distribution and tolerability of risks for social groups, regions, and future generations, leading to the label of "risk society". With such interests running high, environmental problems thus lead to a democratic dilemma for at least two reasons: allocating resources to address such problems requires that we consider possible redistributions of such resources, and there are many sources of relevant knowledge with which to make such decisions (Coenen, Huitema and O'Toole 1998).

Frequently in the past, problem conflicts have been reduced to a difference in interpretation of facts and/or difference in values of stakeholders. The over-simplification leads to a simplistic prescription that conflicts diagnosed to be about facts are in need of "more science", while those about values are in need of "more politics". Such reductionism has dulled the potential of public participation by not recognizing that public value differences are tied to factual uncertainties and trust in public institutions (Renn, Webler and Wiedemann 1995). Recognizing the deficiency in the simple "fact versus value" perspective, Renn, Webler and Wiedemann (1995), expanding on the work of Functowicz and Ravetz (1985), proposed a framework for characterizing problems as a matter of differences at one or more of three levels of debate. The debate levels differ over a range of intensity of problem conflict and degree of problem complexity (see Figure 2.2). The levels of conflict in a problem are due to differences as a matter of worldviews and values, experience and trust, and knowledge and expertise (Renn, Webler and Wiedemann 1995).

A third aspect of social-institutional influence involves the *persons, groups, and/or organizations as convenor of participants* (Cowie and O'Toole 1998, Renn, Webler and Wiedemann 1995, Susskind and Field 1996, Wood and Gray 1991). Wood and Gray (1991) view the convenor as a fundamental influence in setting a topic and direction for discussion, and possibly also the party that acts as the facilitator for the effort. The politics of bringing people together is a high art that is praised by most, and recognized as a fundamental reason how and why complex problems get addressed in the way they do. Renn, Webler and Wiedemann (1995) synthesize a set of citizen participation studies from a research workshop showing that different strategies of participation promote different abilities to give different groups voice to discourse, introduce various types of

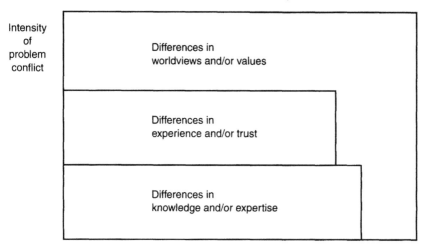

Intensity
of
problem
conflict

Differences in
worldviews and/or values

Differences in
experience and/or trust

Differences in
knowledge and/or expertise

Degree of problem complexity

*Figure 2.2* Three levels of conflict in decision problems based on intensity of conflict and degree of problem complexity (adapted from Renn, Webler and Wiedemann 1995, p. 356)

discourse that are allowed, and that fairness in giving voice differs dramatically (see Table 2.3). Because organizations prefer various types of participation due to familiarity with past experiences, some complex problems are likely to be addressed in standard ways, when an alternative strategy might have worked better (Renn, Webler and Wiedemann 1995, Stern and Fineberg 1996), and perhaps avoiding an opportunity to use a strategy that has continually demonstrated win-win success (Susskind, Levy and Thomas-Larmer 2000).

A fourth aspect of social-institutional influence is *choosing the number, type and diversity of participants* that are brought together to address a problem. Access to the discourse in terms of giving voice to all groups who are affected by a complex problem situation sets up an ironic dilemma. Most research recognizes that the larger the group with different interests being convened, the more opportunity for conflict (Stern and Fineberg 1996). Thus, it is perceived that more interests means longer solution times. However, when all interests are not convened at the beginning of a process, there is more opportunity for challenge, perhaps in the form of lawsuit, at process completion (Renn, Webler and Wiedemann 1995). Thus, even though convening more groups in the short-run might appear to extend a process, not convening the appropriate groups sets up the risk of having the deliberation fail through continual challenges (Susskind and Field 1996). More evidence is building that effective environmental decision processes incorporate broader participation, avoiding the short-term misperception that avoiding conflict is good for the long-term as well

*Table 2.3* Citizen participation strategies in environmental discourse

| | |
|---|---|
| 1. Citizen advisory | people are selected by an institutional body based on representation of major interest positions, but not the full range of interests for logistical reasons |
| 2. Regulatory negotiation | people are selected by the regulatory agency convening the event; as the agency selects the participants, defines the charge, and hires a facilitator |
| 3. Compensation and benefit sharing | a process whereby a group is formed to devise an equitable outcome through developing compensation packages |
| 4. Mediation | people are selected based on interest positions that have enough political clout to interfere with decision implementation |
| 5. Citizen juries | people are selected from a random pool of citizens to evaluate policy alternatives |
| 6. Planning cells | people are selected as a representative set of citizens to foster dialogue and reflection about a specific policy or decision problem |
| 7. Citizen initiatives | people are selected on the basis of those who have been shown by some analysis means to be affected by the decision |
| 8. Dutch study group | people are selected on the basis of showing interest in the problem |

(Coenen, Huitema and O'Toole 1998). This aspect introduces a level of analysis for a decision situation expressed in terms of intra-organization, organization-wide or inter-organizational levels of group decision making (DeSanctis and Poole 1994, Kunreuther *et al.* 1983, Ostrom 1992, Reagan and Rohrbaugh 1990, Roberts and Bradley 1991, Wood and Gray 1991, Raiffa 1982, Stern and Fineberg 1996). Bringing this aspect to central focus allows us to understand the potential differences among groups at different levels, although they might be treating the same subject matter. Although linkages are clearly evident among the levels, little has been done to explore cross-level linkages in a systematic manner. Given that this is the case, it is important that a theoretical approach address multiple levels of decision making. The multiple levels synthesized from the literature (Gray and Wood 1991, Pasquero 1991, Wood and Gray 1991) are: individual, intra-organizational, organization-wide, inter-organizational (supra-organizational community), and society. Although decision making within organizations, i.e. intra-organizational groups and inter-departmental organization-wide, is perhaps the most common context for collaboration (Poole 1985, Campbell and Masser 1995), Wood and Gray (1991) contend that inter-organizational group decision making among autonomous stakeholders is more complex. The difference is that participants from a single organization are already more likely to share a set of organizational rules, norms and structures (Campbell 1996, Sackmann 1991), whereas participants from different organizations might or might not depending on the nature of the issues at hand (Wood and Gray 1991).

For example, Cowie and O'Toole (1998) suggest that it is perhaps more important to link and integrate levels of government in decision making than simply to find the right level to address a problem to make the problem administratively easier. They examine interstate river basin management watersheds that do not pay attention to administrative jurisdictions.

A fifth aspect of social-institutional influence comes from a collection of *rules and norms as social structures among participants*. Such structuring influences interaction in a group by way of modes of participation adopted as a matter of expectations for communication (DeSanctis and Poole 1994, Habermas 1984, Healey 1995, Kunreuther *et al.* 1983, Pinto and Azad 1994, Raiffa 1982, Renn, Webler and Wiedemann 1995, Wood and Gray 1991). Although organizational rules and norms have been a topic of investigation for GIS adoption and implementation (Campbell 1996, Pinto and Azad 1994), we know very little as to how rules and norms impact on participatory GIS use in light of different strategies for participation. Establishing the character of discussion arenas is an aspect of structuring that sets a stage for open or closed settings, including private and public discourse settings (Habermas 1984, Healey 1995, Renn, Webler and Wiedemann 1995, Wood and Gray 1991). Habermas (1984) and Healey (1995) distinguish communicative arenas from strategic arenas, whereby public meetings that follow a presentation-style format are actually considered "intellectually closed" meetings, and public meetings that follow a discussion format are "intellectually open" meetings. Renn, Webler and Wiedemann (1995) describe eight participation strategies in the context of discourse on environmental problems, and recognize that the strategies provide different levels of participation fairness because of the inherent agenda setting associated with the strategies. It is these types of participation implemented through agenda setting that have been of considerable interest to software designers attempting to enhance the way people communicate, cooperate, coordinate, and collaborate with one another over computer networks.

*Construct 2: group participant influence*    One aspect of group participant influence deals with *participants' expectations based on values, goals, issues, beliefs, and fairness* that set the stage for stakeholder perspectives about expected benefits and outcomes (Couclelis and Monmonier 1995, DeSanctis and Poole 1994, Habermas 1984, Kunreuther *et al.* 1983, McGrath 1984, Obermeyer and Pinto 1994, Ostrom 1992, Raiffa 1982, Reagan and Rohrbaugh 1990, Susskind and Field 1996, Wood and Gray 1991). Habermas' (1984) perspectives on values in public planning has influenced researchers in GIS and spatial planning (Couclelis and Monmonier 1995, Healey 1995, Obermeyer and Pinto 1994) to differentiate between the facts that get stored in a GIS and the (social) values used to interpret the facts. Recent theoretical work about common, pooled resources shows that valuing those resources in different ways can lead to different expectations for outcomes (Ostrom 1992). In the context of small groups, competing values have been shown to lead to different interpretations of effectiveness in decision making (Reagan and Rohrbaugh 1990). In a study of consensual approaches to resolving public disputes, Susskind and Cruikshank (1987) identified fairness, efficiency, wisdom, and stability as the four principle goals of a successful dispute resolution process. These goals are complementary, and not mutually exclusive. Continuing to refine the principled negotiation approach into what we call their "participatory negotiation theory", Susskind and Field (1996) identified six basic principles, two of which focus on convening aspects, these being the acknowledgment of concerns from all sides and accepting responsibility, and admitting mistakes and sharing power. Linnerooth-Bayer and Löfstedt (1996), Davy (1996), and Kunreuther, Slovic and MacGregor (1996), and Lober (1995) all suggest that fairness about spatial distribution of impacts from a hazardous facility can have at least three different interpretations: fairness in terms of maximizing the happiness of those being affected overall, fairness in terms of maximizing liberty to those who would locate, and fairness in terms of minimizing the pain of those most disadvantaged. Lober (1995) shows that the "best" location for a hazardous facility would be different depending on which assumption about fairness is accepted. Unfortunately, interpretations about fairness are often not made clear until participants are embroiled in controversy (Linnerooth-Bayer and Löfstedt 1996).

As a second aspect of participant influence, *participants' views/knowledge of the subject domain and each other* mainly involve how participants approach the importance of the topic and how they approach each other in terms of "friendship" or "enemy" feelings (DeSanctis and Poole 1994, Habermas 1984, Kahneman 1992, Raiffa 1982, Susskind and Field 1996, Wood and Gray 1991). Stakeholder views develop as a result of experience and educational background with topics and one another, such that people build for themselves a frame of reference for particular issues, and they sometimes share them. Frames of reference are anchored by "reference points", i.e. familiar information elements that are used as a basis for interpreting material and each other's backgrounds (Kahneman 1992). Often, it is because of the similarities and differences among people in their worldviews and values, experience and trust, and knowledge and expertise that people align themselves into stakeholder groups; worldviews and values being important to the alignment, with experience and trust next, and knowledge and expertise less important (see Figure 2.2). The difference in stakeholder perspective leads to different values, objectives and criteria being articulated as the basis of solutions of group-influenced problems (Keeney 1992). When values, objectives and criteria do not vary significantly among people, then we can say that the participatory decision process is likely to be much like that for a single individual, as we

do not expect (most) single individuals to be internally (cognitively) con-
flicted with regard to interests and facts in problems. However, the non-
conflictual circumstance seems to come about less frequently in
public-oriented problems these days than does the conflictual circumstance.

A third aspect concerns *participants' trust in the process* (Kunreuther,
Slovic and MacGregor 1996, Susskind and Field 1996). Kunreuther and his
colleagues, among many others, have observed an erosion of trust in
public processes when multiple citizen perspectives are not considered.
The erosion of trust is linked to a disenchantment on the part of the public
with representative democracy, i.e. the failure of periodic elections to indicate sufficiently the wishes of voters and from apparent failure of
representative institutions to deal with long-term problems that may not
immediately influence the outcomes of elections. Both of those factors are
combined with the closing of a gap in education levels between representatives and voters compared to the founding days of modern democratic
systems. Because many publics are disillusioned with the political process,
participation in collaborative decision efforts is increasing (Coenen,
Huitema and O'Toole 1998, Duffy, Roseland and Gunton 1996). Furthermore, the increase in non-governmental organizations (NGOs) is seen as a
failure of government to hold a sustained interest in certain problems,
such that NGOs can fill an institutional gap for organizing the interests of
the public — keeping pressure on elected officials from one administration
to the next.

A fourth aspect of participant influence comes from *participants' beliefs
and feelings about technology* (DeSanctis 1993, Shiffer 1998, Turk 1993).
Turk (1993) recognizes that emotive issues for GIS technology are under-
studied, and in some cases may be as important as the technology itself.
Feelings and beliefs are likely to be an important aspect of reinforcing a
person's experience with technology (DeSanctis and Poole 1994). Such
experience encourage and/or hinder expressing one's interest in considering
new ways of accomplishing tasks (DeSanctis and Poole 1994, Shiffer 1998).

*Construct 3: participatory GIS influence*   One aspect of social-technical
information structuring deals with the combination of *place, time, and
channel of communications.* Whereas the rules and norms for social structuring are a social-institutional aspect of participation, different types of meetings structured in terms of place, time, and communication channels also
have an impact on who says what and when during participation in a
decision situation (Habermas 1984, Healey 1995, Jarke 1986, Nyerges 1995,
Webler and Wiedemann 1995, Wood and Gray 1991). The physical (or
virtual setting) of a place has a significant impact on whether people attend
a discussion. Being able to attend a meeting due to scheduling (distance and
timing) constraints is a fundamental concern in participation (Renn, Webler
and Wiedemann 1995). More local and more frequent meetings do not

always enhance the opportunity to participate, since more time away from some other (perhaps work) activity is not always as convenient. It is for that reason that technology supported meetings have been on the increase to open channels of communication. In work related to such efforts, four types of meeting environments are feasible, each described in terms of place and time, which tend to constrain and/or foster human-computer-human interaction (Jarke 1986, Nyerges 1995). The most common type of meeting, called a face-to-face meeting, supports a group convened in the same place at the same time, whereby work can be performed in a conference room. A computer might or might not be used, but often large screen projectors are very helpful to visualize displays of information. A second type of meeting, called a storyboard meeting, involves the same place but different time, whereby work is performed by using leave-behind materials, perhaps supported by a local or wide area computer network. A third type of meeting, called computer conferencing meeting, involves different places but the same time, whereby work can be performed using interactive desktop audio and video, supported by a wide area network, a dedicated wide band-width telephone line, or a satellite link. A fourth type of meeting, called a distributed meeting, involves different places and different times, whereby work can be performed using e-mail, wide area network, and network-resident multimedia tools. Each of the four types of meetings and the interaction that each encourages can be best supported with different computer network configurations.

A second aspect of social-technical information influence involves the *availability of social-technical structures as information aids* (Armstrong 1993, Calkins and Obermeyer 1991, Campbell and Masser 1995, DeSanctis and Poole 1994, Dickinson 1990, Kunreuther *et al.* 1983, Harris and Weiner 1998, Healey 1995, McGrath 1990, Obermeyer 1998, Obermeyer and Pinto 1994, Pinto and Azad 1994, Shiffer 1998, Susskind and Cruikshank 1987, van der Schans 1990). These structures provide information aids to support the participatory effort, e.g. maps, tables, diagrams, sound, and/or spreadsheets. Such aids are often thought of as the tools. For some practitioners, spatial decision support is synonymous with user-friendly and flexible access to decision-relevant data, stored in a spatially-indexed, GIS database (Harris and Weiner 1998, Obermeyer 1998). Indeed, some tactical spatial decisions (i.e. allocation of forest fire fighting crews and equipment) can be fully supported by the results of GIS database query. For others, spatial decision support involves an ability to perform "deep" thinking (e.g. evaluation and interpretation) about a complex spatial problem, in an interactive and iterative manner, such that the decision maker(s) is supported in proceeding toward some single conclusion, or a series of conclusions (Moore 1997, Shiffer 1998). In this latter view, three formerly separate types of decision support are integrated into a spatial decision support system. These are cartographic visualization tools, spatial query tools, and analytical models. Computer mapping techniques implement cartographic visualization tools. Spatially referenced database management systems implement spatial query tools. Spatial analysis techniques support analytical model development and decision analysis techniques make use of the results of spatial analysis to introduce evaluation of multiple alternatives for decision making. A spatial decision support system integrates those techniques in a computerized, analytical environment that supports decision-makers in their search for solutions. In addition to those capabilities one can add nuances to visualization that support consideration of multiple perspectives on decision problems. To extend social-technical information support for groups one can add group communications technology that includes capabilities for agenda setting as described above in the social-institutional construct. In Chapter 3 we elaborate on our presentation of these multiple methods and techniques that could be made available to compose a participatory GIS (PGIS) for decision support. A PGIS is meant to support the explicative (for everyday conversation), technical (for detail of specific disciplinary concerns), practical (for supporting social interaction), and therapeutic (for authenticity) concerns in discourse associated with communicative action (Habermas 1984).

### 2.3.1.2 Constructs and aspects of participatory process as social interaction

The matrix in Table 2.1 represents an expanded view of a participatory decision process flowing from appropriation (construct 4), decision process (task) management (construct 5), and emerging information (construct 6), and back to appropriation and round again. These three constructs consist of six aspects.

*Construct 4: appropriation*   Appropriation is the act of invoking a structure, whether the act is one-time or continual (DeSanctis and Poole 1994). Continual appropriation of the same structure can be called "use" but does not include the act of continual use once invoked. Various appropriation acts structure the character of the information use. One aspect of appropriation involves *social-institutional influences*, i.e. appropriating at any time any one or more of the five aspects of social-institutional influences (DeSanctis and Poole 1994, Gray 1989, Habermas 1984, Healey 1995, Kunreuther *et al.* 1983, McGrath 1984, Pinto and Azad 1994, Raiffa 1982, Susskind and Field 1996, Wood and Gray 1991). Institutional mandates and authority control the flow of information within group situations (Gray 1989, Kunreuther *et al.* 1983, 1996, Susskind and Field 1996). Organizations can be expected to follow their mandates when they are faced with shared problems (Ostrom 1992). Organization-specific and inter-organizational mandates might at times be at odds, thus appropriation can change the way representatives of organizations posture their interests. When people come together in collaborative alliances the topic that brought them together is used to "keep the participants on track", although from time to time other significant concerns arise that might or might not be treated (DeSanctis and Poole 1994, Healey 1995, Kunreuther *et al.* 1983, McGrath 1984, Raiffa 1982, Wood and Gray 1991). Based on agreement among people and/or organizations a convenor often identifies a facilitator to assist with internal group process. Participants can appropriate that control when they ask a facilitator to referee discussions from time to time. From the mandates and the task at hand that brings people together, facilitators help convenors establish agendas (both meeting agendas and project agendas) to help groups organize themselves (DeSanctis and Poole 1994, Habermas 1984, Healey 1995, Kunreuther *et al.* 1983, Pinto and Azad 1994, Raiffa 1982, Wood and Gray 1991). Such agendas might form the basis of the macro-phases and the micro-phases of decision processes in the long and short-term, respectively. Whether participants adhere to such agendas is a matter of what transpires during analytical-deliberative debate involving the information brought to light. Participants have a way of changing agendas to suit expression of their interests.

A second aspect of appropriation concerns *appropriation of group participant influence*. When participants are recognized by others in the participation process then they provide voice to certain concerns (DeSanctis and Poole 1994, Kunreuther *et al.* 1983, Stern and Fineberg 1996, Raiffa 1982, Roberts and Bradley 1991, Wood and Gray 1991). These concerns might be important to interested and affected parties, the concerns of clarification from technical specialists and/or concerns of allocating a redistribution of resources by managers or decision makers (Stern and Fineberg 1996). Participants as the "agents of change" in conversation introduce information about concerns based on their trust with the process of getting a "fair voice". Sometimes that information is introduced through the use of advanced information technology based on their belief that such technology treats information in a way that suits their need for information (DeSanctis 1993, Turk 1993).

A third aspect of appropriation deals with *appropriation of participatory GIS influence*. We know little about place, time and communication channel influences on information use in geographic problem solving and decision making. Most people recommend anytime, anywhere access, as this covers all situations. However, just what advantage and disadvantage there is to the various forms of place and time meetings when map information is being discussed is not known. In addition to physical settings, we know little about how information technology can be best put to use under different types of participatory process methods, such as those identified by Renn, Webler and Wiedemann (1995). Thus, there is still much opportunity for exploration and research. However, it is fairly clear that a deepened problem understanding can be fostered by exposure to information from a broader array of perspectives, as well as by more detailed information that is made more accessible through graphics. To foster that understanding, powerful, group-oriented decision support tools are needed (Armstrong 1993, Jankowski *et al.* 1997). However, there is an inherent trade-off between the sophistication (representational and analytical power) and ease of use of decision support tools. More sophisticated tools are often more challenging to use, despite the enormous effort put into making tools "user friendly". This tradeoff becomes even more significant when the users of decision support tools are people of various educational and cultural backgrounds. A non-specialist approaching GIS software often naively expects to work with a set of virtual maps that portray an "objective, shared understanding" about the world. Yet, since GIS has roots in many disciplines its effective use requires a considerable knowledge, suggesting that there are "multiple realities" to be portrayed.

The appeal of using GIS to support the participatory decision making process comes from the finding that on average, people can understand graphics more easily than tables for many types of problem. An image, a drawing, a map convey information more succinctly, if not better, than a table of numbers, a textual description, or a mathematical equation (Jarvenpaa 1989, Jarvenpaa and Dickson 1988). Yet, since GIS integrates spatially referenced data with analytical functions, some researchers have criticized it as a construction of positivist thinking constraining alternative views of reality that otherwise might broaden the decision making discourse (Lake 1993). Others, on the contrary, argued for more analytical capabilities and decision support functions (Densham 1991).

Any participatory approach to decision making is an organic process that progresses from an open discourse through analysis, argumentation, and negotiation, as was portrayed by the macro-micro strategy outlined in section 2.2. A typical spatial decision problem that lends itself to a participatory approach has a mix of unstructured and structured components, and may involve cooperative, coordinated and/or collaborative decision activities. GIS technology is quite capable of dealing with structured spatial decision problems, but less capable of tackling unstructured parts of a problem, which is why some research in spatial decision support system continues to focus on the process of interaction. Some progress in the development of support tools for participatory spatial decision making is already evident. The development of tools has proceeded along three forms defined by different arrangements of space and time. In the same-place/same-time meeting the technical aspects of using GIS technology are handled by a GIS analyst who also plays the role of human chauffeur (someone who drives the operation of the system), while the agenda and organizational aspects are handled by a meeting facilitator (Faber *et al.* 1995, Jankowski *et al.* 1997). The other two arrangements focus on same-place/different-time, and different-place/different-time meetings (Jankowski and Stasik, 1997). They pose technological challenges in communication management and methodological challenges in developing automated capabilities substituting for human expertise. They also require tools that allow non-specialist users to perform analytical tasks equal to those performed by specialists.

*Construct 5: group process*   One aspect of group process concerns *idea exchange as social interaction* (Campbell and Masser 1995, DeSanctis and Poole 1994, Habermas 1984, Healey 1995, Kunreuther *et al*. 1983, McGrath 1990, 1991, Obermeyer 1998, Raiffa 1982, Reagan and Rohrbaugh 1990, Renn, Webler and Wiedemann 1995, Roberts and Bradley 1991, Stern and Fineberg 1996, Susskind and Cruikshank 1987, Susskind and Field 1996, Wood and Gray 1991). The process concerns the kind of ideas about issues, the way they are brought into the conversation and the way these ideas influence the direction of decision making. Several researchers have developed various ways of describing group process that are relevant for geographic decision situations. Each of these approaches could be used as a basis of developing coding schemes for empirical research similar to those presented by Nyerges *et al*. (1998).

Renn, Webler and Wiedemann (1995) used the concepts of access, communicative competence and fairness presented by Habermas (1984) to evaluate eight participation strategies as related to environmental discourse (see Table 2.4). We adapted their evaluation, portraying each strategy in terms of rank numbers for characteristics of the three concepts. In terms of "access to voicing an idea" access to discourse can be highly restrictive in the case of citizen advisory committees (rated with a 1) or very open in the case of Dutch study groups (rated with a 7). Whether or not any particular macro-micro activity-phase in a project, e.g. as per Table 2.1 in section 2.2, is characteristic of that level of restriction is a matter of the particular project and meeting.

Renn, Webler and Wiedemann (1995) used Habermas' concept of communicative competence to characterize the ability of participation strategies to support different types of discourse, providing participants with an ability to share ideas through styles of language. Four types of discourse were characterized: explicative, theoretical, practical, and therapeutic (with a rating of 1 indicating little ability and a rating of 7 indicating significant ability). The explicative discourse involves terms, definitions, grammar and the everyday use of language. Group process should allow conversations that make reference to worldly events in everyday language. The theoretical discourse involves references to scientific studies as in an objectified world. Group processes should also allow reference to detail of the nuances of complex problems described in terms of technical (discipline-based) languages. The practical discourse involves social needs and the appropriate forms (norms) of social interaction. Group processes must support social interaction that develops out of conventions people know from their everyday experience. The therapeutic discourse makes reference to the subjectivity of a speaker in terms of sincerity and authenticity of claims. Conflict is undoubtedly going to arise in complex situations

Table 2.4 Participatory strategies in terms of access, competence and fairness of discourse (See text for explanation of cell ratings)

| Strategy | Access | Communicative competence | | | | Fairness | | | Rank sum |
|---|---|---|---|---|---|---|---|---|---|
| | | Explicative | Theoretical | Practical | Therapeutic | Agenda | Moderator | Discussion | |
| Planning cells | 4 | 4 | 5 | 7 | 4 | 7 | 4 | 7 | 42 |
| Mediation | 3 | 7 | 4 | 4 | 1 | 7 | 7 | 7 | 40 |
| Citizen initiatives | 5 | 7 | 1 | 5 | 4 | 4 | 4 | 7 | 37 |
| Dutch study group | 7 | 7 | 7 | 4 | 2 | 1 | 1 | 7 | 36 |
| Citizen advisory | 1 | 7 | 3 | 3 | 4 | 4 | 7 | 5 | 34 |
| Regulatory negotiation | 1 | 7 | 7 | 1 | 1 | 1 | 1 | 7 | 26 |
| Citizen juries | 4 | 7 | 2 | 4 | 1 | 1 | 3 | 4 | 26 |
| Compensation and benefit sharing | 2 | 1 | 4 | 5 | 1 | 4 | 1 | 1 | 19 |

where values differ. Coping with those conflicts can be handled by permitting therapeutic discourse. From Table 2.4 one can see that no single participation strategy works best across all competencies. How these strategies encourage certain types of group process for certain types of problem tasks, effecting certain types of outcomes, is still a matter of empirical research.

Stern and Fineberg (1996) led a US National Research Council commission in a discussion about the social-behavioral, group process involved with understanding risk in complex, decision situations. They synthesized a process called an "analytic–deliberative decision approach". The process mixes components of deliberative conversation with analytical investigations into risk-prone situations. These two characteristics of risk-prone situations could be used to describe group process in risk-prone decision situations, if suitable coding schemes could be developed similar to those for decision functions presented by Nyerges *et al.* (1998). Each characteristic could be unpacked into respect aspects of deliberation steps and analytical steps and used to detail the character of group process. Not surprisingly, GIS was mentioned as an information technology that could be useful in several risk-prone decision situations.

Effective decision process is the topic treated by competing values theory (Reagan and Rohrbaugh 1990) that can be used to describe group process. Four perspectives on effective decision processes are described using three dimensions. The perspectives are consensual, empirical, political, and rational. The dimensions are "flexibility or control" of process, "internal or external" grounded focus to group activity, and "process (means) criteria versus results (ends) criteria" for orientation. Each perspective engenders a particular collection of values to make that process a worthwhile endeavor. A consensual perspective promotes flexibility and internal focus, with its effectiveness criteria being participatory process and supportability of decision. The values engendered with such a perspective are dialectical-conflictual, existential, sympathetic and feelings and social compromise. An empirical perspective promotes control and internal grounded focus, with its effectiveness criteria being data-based process and accountability of decision. The values engendered with such a perspective are inductive-deductive, empirical, matter-of-fact, and information utilization. A political perspective promotes flexibility and external grounded focus, with its effectiveness criteria being adaptable process and legitimacy of decision. The values engendered are synthetically-representational, interpretive, insightful, and realism and resources. A rational perspective promotes control and externally grounded focus with its effectiveness criteria being goal-centered process and efficiency of decision. The values engendered are formal-deductive, rational, logical, subjective rationality. These perspectives are not mutually exclusive with regard to a particular decision strategy, but some of each perspective might be found. Codes for these could also be constructed to describe group process in a similar manner to communicative competence and analytic-deliberative processes.

A second aspect of group process focuses on *participatory task flow management* which concerns the structuring into stages, steps or phases, either from a pre-determined agenda or an open agenda, or mixture of both (DeSanctis and Poole 1994, Gray 1989, Healey 1995, Kunreuther *et al.* 1983, McGrath 1990, 1991, Raiffa 1982, Susskind and Cruikshank 1987). When a group monitors its own process it is more likely to stay on track (Poole and Roth 1989). Renn, Webler and Wiedemann (1995) characterized the fairness of participation strategies in terms of three criteria: agenda and rule making, moderation and rule enforcement, and discussion (with a rating of 1 indicating a low level ability of the model, and a rating of 7 indicating a high ability of the strategy), as provided in Table 2.4. Agenda and rule making deal with who participates in setting the agenda and the rules by which the group will interact. Moderation and rule enforcement deal with whether the group process is facilitated and whether a facilitator enforces the rules that have been established. Discussion involves the degree to which all who are affected by the decision have a voice in the process. Group process that is fair is a basic right in a direct democracy. Few complex public–private problems are addressed through direct democracy, more of them are addressed through a representative democratic process. However, one of the major issues seems to be that representative democracy is not functioning as it once did (Coenen, Huitema and O'Toole 1998). Thus, participatory decision situations appear to be on the rise.

Participation strategies contextualize each of the macro-phases of group process by way of setting agendas for decision situations. However, this is not the same as a decision strategy described in section 2.2. A decision strategy is akin to a project agenda, rather than a meeting agenda. The example decision strategy following that reported in Renn *et al.* (1993), and presented in Table 2.1 consists of developing values and criteria, designing a set of options that provide potential solutions, and evaluating those options to recommend a choice. For each of those macro-phase tasks a task agenda could be set to accomplish a task. As discussed in section 2.2, Bhargarva, Krishnan and Whinston (1994) used Simon's (1977) decision steps of gathering, organizing, selecting and reviewing information to describe how a micro-decision process is at work for every macro-phase in a decision. These phase-activity steps provide a "general" representation of what must be done. However, every group sets its own agenda for performing work, and we would not expect anything different here. Thus, these general phase-activity steps represent a normative approach that could be used in empirical research for comparing phase-activity sequences to what actually transpires. The coding systems referred to above could be developed in a manner similar to those described by Nyerges *et al.* (1998).

A third aspect of group process is the *behavior of participants toward each other*. This concerns the working relationships that develop as ideas are exchanged and the decision process proceeds (DeSanctis and Poole 1994, Habermas 1984, Healey 1995, Kunreuther *et al.* 1983, Raiffa 1982, Susskind and Cruikshank 1987, Susskind and Field 1996, Wood and Gray 1991). Stakeholder behavior involving conflict has been studied, and both idea differentiation and integration are important (DeSanctis and Poole 1994, Susskind and Cruikshank 1987). Obermeyer and Pinto (1994) see the introduction of GIS as encouraging more conflict between groups rather than less. Nyerges and Jankowski (2001) investigated group conflict in decision support for habitat redevelopment site selection, and found that maps are less likely to be associated with discussion conflict than tables — tables being more analytic displays for priority ranking of habitat sites than are maps for ranking the same sites.

*Construct 6: emergent influence*   One aspect of this construct is the *emergence of social-technical information influence* (Calkins and Obermeyer 1991, DeSanctis and Poole 1994, Dickinson 1990, Healey 1995, Kunreuther *et al.* 1983, McGrath 1990, 1991, Obermeyer and Pinto 1994, van der Schans 1990). Although various technological capabilities are provided by software and hardware according to the design of a system, certain other emerging structures might come to light during the treatment of information. The emergence of social-technical structures, such as new map designs or database designs, might help a group with further information structuring. However, such emergence could make information easier or more difficult to understand in the longer term. Consequently, the emergence of social-technical information structures has a rather significant impact on what information a group treats from activity-phase to activity-phase (DeSanctis and Poole 1994).

A second aspect of this construct is the *emergence of group participant influence* (DeSanctis and Poole 1994, Healey 1995, Renn, Webler and Wiedemann 1995, Roberts and Bradley 1991, Wood and Gray 1991). A better understanding of values, goals, objectives, and beliefs are bound to come to light through participant conversation (Renn, Webler and Wiedemann 1995, Wood and Gray 1991). Participants might clarify their own perspectives and/or the perspectives of others in regards to values, goals, objectives, and beliefs. Participants might clarify or muddle their views of subject domain as they converse with others. Views of each other in regards to respectful opinion of what others have to say will undoubtedly get refined (Wood and Gray 1991). Trust in each other might change as a result of ideas being exchanged. Those who encourage use of technology can have an impact on the feelings that participants develop regard to its continual use. Feelings for people and technology might well be connected.

A third aspect of this construct is *emergence of social-institutional influence that* deals with how rules or norms are brought into use and eliminated or reinforced during the decision process (DeSanctis and Poole 1994, Healey 1995, Roberts and Bradley 1991, Wood and Gray 1991). Clarifying mandates and the problem at issue can lead to refocusing activity for any particular task. In regards to agendas, one can choose to make use of rules to keep the conversation on track or to derail it. The emergence of new rules about how people communicate during the participation changes the course of the interaction.

### 2.3.1.3 Constructs and aspects about participatory outcomes

Two constructs are part of the participatory outcomes in EAST2, task outcomes and social outcomes.

*Construct 7: task outcomes* Two aspects appear to be fundamental in regards to task outcomes. One aspect is *character of decision outcome* (Coenen, Huitema and O'Toole 1998, Cowie and O'Toole 1998, DeSanctis and Poole 1994, Dickinson 1990, Healey 1995, McCartt and Rohrbaugh 1995, McGrath 1984, 1990, 1991, Obermeyer and Pinto 1994, Ostrom 1992, Ostrom, Gardner and Walker 1994, Pestman 1998, Pressman and Wildavsky 1984, Reagan and Rohrbaugh 1990). Reports on task outcomes have been provided in early studies of group support systems (Chun and Park 1998, Jessup and Valacich 1993). Because decision outcomes tend to be so diffuse, depending on the circumstances of the collaboration effort, some researchers put little emphasis on it (Wood and Gray 1991, Webler 1995). Consequently, many researchers, including those exploring the use of group support systems, recommend focusing on process rather than outcome (Jessup and Valacich 1993, Renn, Webler and Wiedemann 1995). From a different perspective, Duffy, Roseland and Gunton (1996) make a good point about expectations of outcomes perhaps being different than the actual outcomes, thus it is necessary to know what outcomes exist. Pressman and Wildavsky (1984) have looked at the complex of decision outcomes in Federal Program implementations, characterizing the increments of progress in terms of decision points followed by clearance points. Clearance points are the agreement "buy offs" to proceed to next steps. Unpacking long decision processes as decision chains composed of decision points plus clearance points provides an understanding of the incremental character that intermediate decision outcomes play in a process. More recently, several researchers have begun to look at effectiveness of decisions in real settings (Coenen, Huitema and O'Toole 1998, Cowie and O'Toole 1998) indicating that even Webler (1995) suggests it is a good idea, although not pursued due to absence of a good framework. Pestman (1998) distinguishes improvement in process versus improvement in outcome in environmental problems. Cowie and O'Toole (1998) elaborate on decision making effectiveness as a way to assess quality of decision by saying that the multiple characteristic of decision processes can contribute to effectiveness of environmental decisions.

The second aspect linked to task outcome concerns *decision outcome and participant structuring dependence* (McGrath 1990, 1991, Raiffa 1982, Susskind and Cruikshank 1987, Susskind and Field 1996, Wood and Gray 1991). The major concern is the stability of an outcome based on whether it lasts beyond the group that has been formed (Susskind and Field 1996, Wood and Gray 1991). Several researchers suggest that "decision sustainability" is a pragmatic, substantive criterion that could be used to evaluate the quality of decision outcomes (Burns and Ueberhorst 1988, Coenen, Huitema and O'Toole 1998). Decision sustainability involves the ability to manage worldly events in such a way as to preserve the "validity" of the decision without having to overturn the action that was decided. Decision sustainability can be considered at three levels, paradigmatic, instrumental and concrete, i.e. long-term, medium-term, and short-term respectively.

*Construct 8: social outcomes*   One aspect that concerns social-institutional outcomes deals with whether there is an *opportunity for challenge of the outcome* (Habermas 1984, Healey 1995, Renn, Webler, and Wiedemann 1995, Stern and Fineberg 1996, Susskind and Field 1996). The degree to which any decision issue is final and whether it can be changed through further considerations is rather important to promote the results of a process (Habermas 1984, Healey 1995). Certain of the participation processes facilitate the opportunity for challenge, whereas others prohibit it outright (Renn, Webler and Wiedemann 1995). Risky decisions should always be amenable to challenge if new and better information arises (Stern and Fineberg 1996). Each of these challenges could be considered a clearance point to proceed to the next phase-activity in the same way as suggested for decision points by Pressman and Wildavsky (1984).

A second aspect of this construct concerns the *reproduction and temporality of the group participant structuring* (DeSanctis and Poole 1994, Wood and Gray 1991, Habermas 1984, McGrath 1990, 1991, Roberts and Bradley 1991, Susskind and Cruikshank 1987, Susskind and Field 1996). This aspect deals with the stability and longevity of the social relationships among group participants, particularly as promoted in multiple meetings (DeSanctis and Poole 1994). Successful decision situation results from phase to phase encourage and solidify working relationships (Wood and Gray 1991). Pestman (1998) provides an example that shows that participatory contributions can spread through decision processes far beyond the immediate and direct. This is one place where the possibility of participation-based learning is introduced, a theme also developed in parallel with Pretty (1998). Such subtle decision making impacts would seem to be a fertile subject for more in-depth research (Coenen, Huitema and O'Toole 1998).

A third aspect of this construct concerns the *reproduction and temporality of social-institutional structuring* (DeSanctis and Poole 1994, Habermas 1984, McGrath 1990, 1991, Ostrom 1992, Ostrom, Gardner and Walker 1994, Roberts and Bradley 1991, Susskind and Cruikshank 1987, Susskind and Field 1996, Wood and Gray 1991). Such structuring in regards to changing mandates for power or control is only likely to occur over repeated projects — either successes or failures. However, continued successes with projects encourage similar projects to be addressed in the same way, whether this involves the task domain and/or the way that participants are convened.

Eight constructs with the respective 25 aspects presented above likely constitute the single largest enumeration of issues concerning decision making within a geographic context. In addition, each aspect can spawn several variables when making the EAST2 operational, contributing to the comprehensive nature of EAST2. The above constructs and aspects together with respective variables are but concepts that in and of themselves can be used only for developing a "task description" of each macrophase in a decision situation. It is the premises connecting these constructs, aspects and variables that turn a conceptual framework into a theory.

### 2.3.2 Premises of EAST2

Premises are fundamental statements that tie each aspect on one side of the premise to an aspect of the other side. Premises motivate one or more research questions (as for example in Table 2.5), which, when phrased in terms of variables, can be considered hypotheses (propositions) about the dynamics of a macro-phase in a decision situation. The research questions presented in Table 2.5 are a few examples of questions that could be posed. Remember, each premise is a general statement. Each premise statement consists of a *subject construct* related to an *object construct*. Hence, each research question asks something about how a *subject aspect* relates to an *object aspect,* thus many different questions could be posed. In the context of empirical, social-behavioral research we could say: "how does one variable relate to another variable?" It would be out of place in a theory of structuration to refer always to a specific aspect as either an independent variable or dependent variable. In Chapter 4 we take up the issue of "treatment modes" more generally, of which "independent" and "dependent" are but two of six modes. The three categories of convening, process and outcome premises are simply convenient labels to describe social-behavioral relations corresponding to the way the constructs were categorized. We present the premises according to the three following subsections.

*Table 2.5* Example research questions motivated by premises in Enhanced
Adaptive Structuration Theory

| *Premises* | *Research question motivated by resepective premise* |
| --- | --- |
| *Convening premises* | |
| **Premise 1.** Social-institutional influences affect the appropriation of group participant influences and/or social-technical influences | – In what way does the purpose of a decision task influence the types of geographic information structures (e.g. maps, tables, diagrams) appropriated by the participants?<br>– In what way does the organization that convened the decision situation in combination with the diversity of participants influence the type of group participant structuring? |
| **Premise 2.** Group participant influences affect the appropriation of social-institutional influences and/or social-technical influences | – How do the different perspectives such as those oriented to policy/decision maker, technical/scientific specialist, and interested and affected party, influence the types of geographic information structures appropriated?<br>– What types of social-technical information structures appear to be linked to differences in participant structuring? |
| **Premise 3.** Participatory GIS influences affect the appropriation of social-institutional influences and/or group participant influences | – How does each of four meeting venues influence the generation of information structures (e.g. maps, animations, tables, text narration) useful for understanding spatial criteria that can be processed with a GIS?<br>– How do the social-technical capabilities of software get appropriated across meetings in relation to participants' trust in the group agenda which represents a plan for the process? |
| *Process premises* | |
| **Premise 4.** Appropriation of influences affect the dynamics of social interaction described in terms of group processes | – What types of geographic information structures are appropriated during the different intensities of participation that seem to facilitate an analytic-deliberative process, and which information structures seem to hinder the process?<br>– Appropriation of what types of group participant structures has what type of influence on group process? |
| **Premise 5.** Group processes have an affect on the types of influences that emerge during those processes, and emergent influences affect the appropriation of influences | – What kinds of geographic information structures emerge during the different levels of participation in an analytic-deliberative process?<br>– What emergent structures influence the type of appropriation that is undertaken? |

*Table 2.5* Continued

| Premises | Research question motivated by resepective premise |
|---|---|
| *Outcome premises* | |
| **Premise 6.** Given particular influences being appropriated, if successful appropriation occurs and group processes fit the task, then desired outcomes result | – Given that a group appropriates a particular type of information structure that has been found to be useful in the past, and if the information structure is appropriated during "specific conditions", can we expect the outcome from the process to be satisfactory to all participants? <br> – What structure appropriation under what conditions of group process appear to affect the dependence of the decision outcome on group participant structuring? |
| **Premise 7.** Given particular influences being appropriated, if successful appropriation occurs and group processes fit the task, then reproduction of social-institutional influences result | – In what way are various social-institutional structures together with group participant structures linked to the opportunity to challenge the task outcome? <br> – How do inter-organizational protocols and the social interaction during group process promote or discourage further group work? |

### 2.3.2.1 Premises about the convening aspects of a macro-phase decision task

Three premises (P1, P2, and P3) focus on the aspects of convening a participatory decision situation.

*Premise 1* Social-institutional influences affect the appropriation of group participant influences and/or social-technical influences.

The purpose of a decision task might be set by mandate and/or common goal. Common goals are said to be the principal basis of collaborative efforts in an inter-organizational setting (Wood and Gray 1991). Mandates and/or goals influence the kinds of information thought to be relevant to discussions, and hence influence the types of information structures that are useful in presenting information. Interaction norms take their lead from a variety of participants' organizational settings. The working relationships among participants get clarified over a period of time as each participant borrows from their more familiar organizational-based rules (Susskind and Field 1996, Wood and Gray 1991). As social-institutional influences such as mandates and/or goals are borrowed from various organizations, so are the types of information structures thought to be useful in presenting information about decision activity related to those mandates and/or goals. With the above in mind, suitable research questions might be the following. In what way does the purpose of a decision task influence the types of geographic information structures, e.g. maps, tables, diagrams, appropriated by the participants? In what way does the organization that convened the decision situation in combination with the diversity of participants influence the type of group participant structuring?

*Premise 2*   Group participant influences affect the appropriation of social-institutional influences and/or social-technical influences.

Groups have different characters based on what the participants know and how they view the world. Group perspectives often differ as a result of their motivations, values and interests in subject matters. Over several years of work in an attempt to clarify risk communication within communities, Sandman (1993) identified nine publics concerned about community risk-informed problems: neighbors, concerned citizens, technical experts, media, activists, elected officials, business and industry, local, state and federal government regulators. However, in further clarifying the nature of perspectives on risk-informed, controversial topics, others have suggested that perspectives cluster according to three major subcultures: interested and affected parties, policy/decision makers, and technical specialists (Rejeski 1993, Stern and Fineberg 1996). Perhaps it is these three subcultures that account for the most differentiation in the concerns about risk-informed topics with high stakes and high uncertainties (Rosa 1998). Whether the magic number is three or nine or something different is a matter for further empirical research. Regardless of the number, we expect that certain perspectives promote the understanding of information in different ways. If participants with different perspectives are more likely to encourage the appropriation of different types of geographic information structures for appropriation, why is that the case?

Different kinds of group participant structuring allow for different participation strategies (social interaction) in a group (Renn, Webler and Wiedemann 1995). These differences relate to voice, communicative competence, and fairness. With these findings in mind, appropriate research questions might be as follows. How do the different perspectives such as those oriented to policy/decision maker, technical/scientific specialist, and interested and affected party, influence the types of geographic information structures appropriated? Given that different kinds of group participant structuring, i.e. participation strategies, encourage differences in social interaction, what types of social-technical structures appear to be linked to differences in participant structuring?

*Premise 3*   Participatory GIS influences affect the appropriation of social-institutional influences and/or group participant influences.

A face-to-face meeting is a common venue to support a collaborative effort (Wood and Gray 1991). However, advances in groupware now make it possible for meetings to take place in various venues, not just in traditional face-to-face meetings. Actually, four types of space-time meeting venues are possible: same-place and same-time (traditional face-to-face) meeting, different-place and different-time (distributed) meeting, same-place and different-time (storyboard) meeting, and different-place and same-time (conference call) meeting (Jarke 1986). Certain meeting environments are likely to be more conducive to various types of information structures. Because the latter three are relatively newer and more limited in information transmission, we are still exploring how to make use of them. The reason those venues are likely to become more useful is that they each save time and energy. Consequently, a fundamental question during objectives creation for land planning would be: How do each of four meeting venues influence the generation of number and type of information structures useful for understanding spatial criteria that can be processed with a PGIS?

In groupware technology, the "spirit" of a technology can influence appropriations, hence social interaction (DeSanctis and Poole 1994). Spirit is the "intended character" of technology, i.e. the underlying cultural influence to be fostered, and *not* the basic technical design. The spirit of a social-technical capability might foster communication and/or analysis during social interaction, or hinder them. Groupware technology commonly has a "participatory" and "democratic" spirit embedded in its intended use, but groups might put technology to use in ironic (unintended) ways by allowing only certain participants to make use of particular features, suppressing the participatory aspect. Together the above issues encourage research questions like: How do the different perspectives such as those oriented to policy/decision maker, technical/scientific specialist, and interested and affected party, influence the types of geographic information structures appropriated?

We know that different phases of a decision situation influence the appropriation of geographic information structures, so that maps tend to be exploratory and tables tend to be analytical (Jankowski and Nyerges 2001). Given that different kinds of group participant structuring, i.e. participation strategies, encourage differences in social interaction, a relevant research question might be: What types of social-technical structures as information aids appear to be linked to differences in participant structuring?

### 2.3.2.2 *Premises about the process aspects of a macro-phase decision task*

The convening aspects of a collaborative effort influence the process of social interaction in various ways. We take the event of "appropriation" as a lead into the participatory process of social interaction. Two premises (P4 and P5) are important for understanding the dynamics of participatory social interaction in the context of PGIS use (see Table 2.2).

*Premise 4*   Appropriation of influences affect the dynamics of social inter-action described in terms of group processes.

PGIS is designed to influence how a group interacts. The manner in which PGIS capabilities are appropriated can facilitate group interaction, but it can also hinder interaction. Many public–private processes appear to contain both analytical and deliberative phases (Stern and Fineberg 1996). The empirical character of such phases is just beginning to be explored in detail with regard to PGIS technology (Jankowski and Nyerges 2001). However, more generally in undertaking this type of research it is import-ant to recognize that at least four, cumulative levels of "social interaction", hence HCHI, can be elucidated under the umbrella term of "participa-tion": *communication, cooperation, coordination,* and *collaboration.* At a basic level of participation, people communicate with each other to exchange ideas as a fundamental process of social interaction. In public decision contexts the traditional forum of a public meeting provides for *communicative* interaction — but only at a most basic level, a drawback to such meetings when "truly constructive" comments are desired (Schnei-der, Oppermann and Renn 1998). At the next level of social interaction, building on a set of ideas developed through basic communication can be considered to be *cooperative* interaction. Participants in a cooperative activity each agree to make a contribution that can be exchanged, but each can take the results of the interaction away with them and act on the results as they see fit, with no further interaction required. A *coordinated* interaction is one whereby participants agree to cooperate, but in addition they agree to sequence their cooperative activity for mutual, synergistic gain. A *collaborative* interaction is one whereby the participants in a group agree to work on the same task (or subtask) simultaneously or at least with a shared understanding of a situation in a near-simultaneous manner (Roschelle and Teasley 1995). Working in a collaborative manner, participants create synergy, and each comes away with a synergistic sense of how to undertake decision making. On the basis of knowing that differ-ent intensities of interaction exist, a relevant question would be: What types of geographic information structures are appropriated during the different intensities of participation that seem to facilitate an analytic-deliberative process and which information structures seem to hinder the process?

How groups manage their participatory work is likely a function of what structuring capabilities are available to manage the interaction, both within a non-computer supported environment as well as within a com-puter supported context (Jessup and Valacich 1993, Chung and Park 1998). Part of those capabilities would be directed at meeting management activities, e.g. agenda setting and communication techniques, while other capabilities would focus on group-enabled geographic analysis and display techniques. Based on such findings, relevant questions to ask might be: Appropriation of what types of group participant structures has what type of influence on group process? Does an open process relative to a closed access process have anything to do with the types of information treated during the process?

*Premise 5*   Group processes have an affect on the types of influences that emerge during those processes, and emergent influences affect the appropriation of influences.

GIS technology has always been touted as the foundation of a data integration capability for complex geographic problems (Cowen 1988). GIS data integration plays a major role in information creation. Being able to bring disparate sources of information together from various organizations is seen as a major advantage in the use of PGIS. Emergent information structures stem from this ability to bring information sources together (DeSanctis and Poole 1994). We would expect various types of information structures emerging because of the variety of perspectives that come together in public–private activity. As such, appropriate research questions might be: What kinds of geographic information structures emerge during the different levels of participation in an analytic-deliberative process? What emergent structures influence the type of appropriation that is undertaken?

### 2.3.2.3 Premises about the outcome aspects of a macro-phase decision task

Two premises address outcomes of a participatory decision process. One concerns task outcomes (P6) and another concerns social-institutional outcomes (P7).

*Premise 6*   Given particular influences being appropriated, if successful appropriation occurs and group processes fit the task, then desired outcomes result.

Task outcome is a fundamental concern to many in participatory decision making sessions (DeSanctis and Poole 1994, Healey 1995, McGrath 1984, Obermeyer and Pinto 1994, Jessup and Valacich 1993). Social-institutional influences, group participant influences, and social-technical influences in a particular participatory context affect the nature of task outcomes. Because the influence tends to be so diffuse, depending on the circumstances of the collaboration effort, some researchers do not put much emphasis on the character of outcomes (Wood and Gray 1991). Consequently, many researchers, including those exploring the use of group support systems, recommend focusing on process, rather than outcome (Jessup and Valacich 1993, Rohrbaugh 1989, Todd 1995). However, Duffy, Roseland and Gunton (1996) concerned with land resource outcomes, make a good point about expectations of outcomes perhaps being different than the actual outcomes; it is therefore necessary to know what outcomes were expected versus which actually occur. Furthermore, more recent syntheses of research findings about the use of group decision support systems (GDSS) in field studies as opposed to laboratory settings shows increased decision quality and shortened meeting time when using GDSS as compared to conventional meetings (Chun and Park 1998). They also report on high user satisfaction and enhanced decision confidence independent of prior user experience with a GDSS. Consequently, examining task outcomes is perhaps more relevant when researchers take into consideration the type of research design. For that reason, appropriate research questions might be: Given that a group appropriates a particular type of information structure that has been found to be useful in the past, and if the information structure is appropriated during "specific conditions", can we expect the outcome from the process to be satisfactory to all participants? What structure appropriation under what conditions of group process appear to affect the dependence of the decision outcome on group participant structuring?

*Premise 7*   Given particular influences being appropriated, if successful appropriation occurs and group processes fit the task, then reproduction of social-institutional influences result.

Organizations have an administrative structure that we call "bureaucracy". Bureaucracy supports and/or hinders information flow and interaction. One of the challenges for groups, especially ad hoc interorganizational groups, is that a bureaucracy is always difficult to maintain over long time periods. The reproduction and temporality of the group structuring in a group context deals with the stability and longevity of the social relationships among group participants, particularly as promoted in multiple meetings (DeSanctis and Poole 1994, Susskind and Field 1996, Wood and Gray 1991). One aspect of relationships among inter-organizational groups concerns the way decision tasks are addressed over a long term. Habermas (1984) and Healey (1995) suggest that strategic decisions should be open to challenge. Setting down a decision for a long term, without revisiting the assumptions and reinforcing the reasons why such a decision should be followed, often creates instrumental stagnation in a group (Healey 1995). With temporal structuring and long term openness in mind, suitable research questions might be: In what way are various social-institutional influences together with group participant influences linked to the opportunity to challenge the task outcome? How do inter-organizational protocols and the social interaction during group process promote or discourage further group work?

The seven premises in EAST2 presented above, together with the respective example questions, indicate that a wide variety of interesting, empirical research opportunities exist with regard to PGIS use in participatory settings. In Chapter 3 we outline PGIS methods as part of the technology offerings, then in Chapter 4 turn to social-behavioral research methods that help us study the use of PGIS methods to address substantive decision problems.

## 2.4 Conclusion

Any given theory is but a set of relationships, conditioned by an ontology, and in part dependent on an epistemology that directs our focus about important questions to ask about the world. As an ontology we accept that not all public problems are interpreted as problems by all groups. We do not hide from the assumption that social-physical realities structure our agency as human beings and as human beings we structure social-physical realities. Our position is mid-way between an objectivist and relativist view with regard to what constitutes a decision problem — organizations, communities, and institutions will help in calling a decision situation a decision situation, whether this be jail siting, landfill closure, hazardous waste cleanup and transfer, transportation projects in a growth management context, or watershed and water quality protection.

We have argued that complex decision situations can be addressed by decision strategies that have been tested through practical situations in the field. Whether solutions are possible within such strategies just might depend on who is convened to critique and (re)construct the strategy. For many complex problems, current organizations and institutions are in need of inter-organizational assistance. After all, many problems exist because of the narrow perspective that single organizations have had in the past. The recent trends in coalition building attests to a recognition that broader perspectives spanning the gaps in knowledge are likely necessary to address recommendations and solutions.

The macro-micro decision strategy framework in which EAST2 is embedded can serve as a flexible but workable guide for investigations on a number of levels. The macro-micro framework was devised to allow both flexibility and explanatory power as needed, depending on whether one is interested in task descriptions as a basis for research investigations about single organizations and their application of GIS for decision support as presented in Chapter 5, or explanatory theory as a basis for research about inter-organizational groups as presented in Chapters 6 and 7. Given that macro-micro EAST was developed to address geographic decision situations facing inter-organizational groups, it can be easily pared back to address less complex situations.

At the level of a macro-phase in a decision situation, clearly, theories other than EAST can be used to elucidate the concepts and relationships that motivate questions about PGIS use. Some of those theories could actually be the ones that have been used in the synthesis of the 25 aspects now considered in EAST2. A less comprehensive theory would likely involve fewer aspects, but a less comprehensive theory does not provide as much potential for a broad-based understanding about the complexity of

PGIS use in decision situations. But still, why so many aspects? What aspect(s) might we do away with? What is the advantage? What is the simplest, but most trustworthy of explanations about PGIS use, based on which aspects and which premises, that eventually leads us to direct our energies to dissuade certain PGIS designs and encourage others? Enumerating the aspects of a decision situation from one task to another shows us what is of concern on the basis of a task description. Comparing and contrasting aspects from one task to another supports development of an understanding as to what makes a task different. It is not just one aspect, but perhaps many aspects that differentiate tasks. When only one aspect is different, then perhaps there might be an easier transfer of findings from one macro-phase task to another, hence situation to situation. When many aspects are different from task to task, we have a way of unpacking the complexity of the situation. Then, when we introduce the premises to motivate research questions we can gain insight into why and when an aspect differs from one task to the next, and/or one situation to the next.

When making use of EAST2 we do not start with a "blank slate" for complex decision situations. It is for this reason we have pointedly chosen the term "situation" for decision situation. There are usually many aspects that influence a decision situation, one might say that many of them are contextual, hence contributing to the notion of "complex" in complex decision making. What aspects about a decision situation are the predominant influences is a matter of developing empirical research findings. Having more aspects does not mean that all are influential, but they at least might be considered. Zadeh (1965) suggests that the more one looks at the details of the world (interpreted here to mean situation) the more uncertainty one is bound to discover. Assuming that to be the case, if we left the explanation potential of EAST at a level of abstraction representing the convening, process and outcome categories, would that have reduced the uncertainty about decision making? If we left the level of abstraction at the level of eight constructs would that have reduced the uncertainty? If we chose to pursue some, many, or all of the 25 aspects would that reduce the uncertainty? The reason for pursuing research at the level of 25 aspects (and of course perhaps more) is that we are interested in exploring the uncertainty — we are not afraid of it. Within uncertainty is the potential for new knowledge. The goal here is to understand the character of uncertainty and how the uncertainties among the details combine into uncertainties at higher levels of abstraction (Nyerges 1991), that uncertainty being responsible for "chaos" in decision making.

Through abductive processes (Schank 1998) in research we can try to articulate propositions that stem directly from research questions motivated as refinements of the EAST2 premises. As we flip-flop between deductive propositions and inductive empirical evidence to develop evidence by way of empirical findings, it is possible to develop "practically adequate" interpretations of situations (Sayer 1984). If we do this enough times through systematic treatment of several situations we believe we can piece together knowledge through an argumentative process as we offer up the evidence from empirical findings (see Einsenhart and Borko 1993). Sometimes (and perhaps even often) the evidence will be contradictory, but offering the evidence with reference points to other findings will allow us to shed light on the validity of various findings, thus building a more robust understanding about the complexity of PGIS use. Issues about validity are treated in Chapter 4 in the context of empirical research strategies.

In unpacking the complexity of "PGIS use in Society" (Sheppard 1995), a fundamental concern when devising EAST2 was that it be able to express the relationships between social-technical constructs and social-institutional constructs (Orlikowski 1992). This is our attempt at not only deconstructing the relationship of PGIS in society, but also reconstructing PGIS in society. Here we add to critique (as a deconstruction of situations), a reconstruction of ways to improve PGIS. This reconstruction makes the additional contribution toward what we call participatory geographic information science. Articulating social-behavioral explanations by way of the premises and the research questions they motivate is intended to lead to a "deeper understanding" about what the impacts of software designs are on groups in society. Addressing any one or more questions among such a wide variety of questions is a considerable challenge. The challenge "begs" for a systematic approach for evaluation of empirical social-behavioral research methods so that we can better understand of how empirical results relate to each other in our attempts to build knowledge about the implications of PGIS use. In a social-behavioral research study involving primary data collection, choice of a data strategy follows research question articulation as we describe in some detail in Chapter 4.

Lest the reader think that our concern is only theoretical, we have a concern for *praxis* — combining theory and practice — as demonstrated through the materials presented in Chapter 3. The EAST2 framework depicted in Figure 2.1 is actually useful as a template for task descriptions of complex situations through which PGIS requirements can be elucidated. We have found the framework as a set of constructs and aspects can be used to articulate scenarios for PGIS use. The constructs alone point out the multi-faceted considerations about the PGIS use. Our revision of the constructs and aspects in this chapter should only help us do a better job at elucidating those considerations for PGIS requirements. We turn to such a discussion in Chapter 3, as we present PGIS methods that constitute the fundamentals of PGIS requirements.

# References

Armstrong, M.P. (1993) "Perspectives on the development of group decision support systems for locational problem solving", *International Journal of Geographical Systems*, 1(1): 69–81.

Beck, U. (1992) "From industrial society to the risk society: questions of survival, social structure and ecological enlightenment", *Theory, Culture and Society*, 9: 97–123.

Bhargava, H.K., Krishnan, R. and Whinston, A.D. (1994) "On integrating collaboration and decision analysis techniques", *Journal of Organizational Computing*, 4(3): 297–316.

Burns, T. and Ueberhorst, R. (1988) *Creative Democracy: Systematic Conflict Resolution And Policy-Making In A World Of High Science And Technology*, New York, Praeger.

Calkins, H.W. and Obermeyer, N.J. (1991) "Taxonomy for surveying the use and value of geographical information", *International Journal of Geographic Information Systems*, 5(3): 341–51.

Campbell, H. (1996) "Theoretical perspectives of the diffusion of GIS technologies", in I. Masser, H. Campbell and M. Craglia (eds) *GIS Diffusion*, London, Taylor & Francis: 23–45.

Campbell, H. and Masser, I. (1995) *GIS and Organizations*, London, Taylor & Francis.

Chun, K.J. and Park, H.K., (1998) "Examining the conflicting results of GDSS research", *Information and Management*, 33: 313–25.

Coenen, F.H.J.M, Huitema, D. and O'Toole, L.J. (1998) "Participation and environment", in F.H.J.M Coenen, D. Huitema and L.J. O'Toole (eds) *Participation and the Quality of Environmental Decision Making*, Dordrecht, Kluwer Publishers: 1–20.

Couclelis, H. and Monmonier, M. (1995) "Using SUSS to resolve NIMBY: how spatial understanding support systems can help with the 'not in my back yard' syndrome", *Geographical Systems*, 2: 83–101.

Cowen, D. (1988) "GIS versus CAD versus DBMS: what are the differences?" *Photogrammetric Engineering and Remote Sensing*, 54, 11: 1551–5.

Cowie, G.M. and O'Toole, L.J. (1998) "Linking stakeholder participation and environmental decision making: assessing decision quality for interstate river basin management", in F.H.J.M. Coenen, D. Huitema and L.J. O'Toole (eds) *Participation and the Quality of Environmental Decision Making*, Dordrecht, Kluwer Publishers: 61–70.

Davy, B. (1996) "Fairness as compassion: towards a less unfair facility siting policy", *Risk* 7(2): 99–108.

Densham, P.J. (1991) "Spatial decision support systems", in D.J. Maguire, M.F. Goodchild and D.W. Rhind (eds) *Geographical Information Systems: Principles and Applications*, first edn, New York, John Wiley & Sons: 403–12.

Dewey, J. (1933) *How we Think, a Restatement of the Relation of Reflective Thinking to the Educative Process*, Boston, D.C. Heath and Company.

DeSanctis, G. (1993) "Shifting foundations in group support system research", in L.M. Jessup and J.S. Valacich (eds) *Group Support Systems*, Macmillan, New York: 97–116.

DeSanctis, G. and Poole, M.S. (1994) "Capturing the complexity in advanced technology use: adaptive structuration theory", *Organization Science*, 5:2: 121–47.

Dickinson, H. (1990) *Deriving a Method for Evaluating the Use of Geographic Information in Decision Making*, Ph.D. dissertation, State University of New York at Buffalo, National Center for Geographic Information and Analysis (NCGIA) Technical Report: 90–3.

Drew, C., Nyerges, T., McCarthy, K. and Moore, J. (2000) *Using Decision Chains to Explore Four DOE-EM Decisions: A Cross-Case Analysis*, unpublished manuscript, Consortium for Risk Evaluation with Stakeholder Participation, University of Washington.

Duffy, D.M., Roseland, M. and Gunton, T.I. (1996) "A preliminary assessment of shared decision-making in land use and natural resource planning", *Environments*, 23(2): 1–16.

Eisenhart, M. and Borko, H. (1993) *Designing Classroom Research*, Boston, Allyn and Bacon.

Electrical Power Research Institute (1998). http://www.smartplaces.com.

Faber, B., Wallace, W. and Cuthbertson, J. (1995) "Advances in collaborative GIS for land resource negotiation", *Proceedings, GIS '95*, Ninth Annual Symposium on Geographic Information Systems, Vancouver, BC, March 1995, GIS World, Inc., Vol. 1: 183–9.

Functowicz, S. and Ravetz, J. (1985) "Three types of risk assessment: a methodological analysis", in V. Covello, J. Mumpower, P. Stallen, and V. Uppuluri, *Environmental Impact Assessment, Technology Assessment, and Risk Analysis*, Berlin, Springer-Verlag: 831–48.

Giddens, A. (1984) *The Constitution of Society: Outline of the Theory of Structuration*, Cambridge, UK, Polity Press.

Gray, B. (1989) *Collaborating: Finding Common Ground for Multiparty Problems*, San Francisco, CA, Jossey-Bass Publishers.

Gray, B. and Wood, D.J. (1991) "Collaborative alliances: moving from practice to theory", *Journal of Applied Behavioral Sciences*, 27(1): 3–22.

Habermas, J. (1984) *The Theory of Communicative Action, Volume 1: Reason and the Rationalization of Society*, Cambridge, Polity Press.

Harris, T. and Wenier, D. (1998) "Empowerment, marginalization, and 'community-integrated' GIS", *Cartography and Geographic Information Systems*, 25(2): 67–76.

Healey, P. (1995) "The communicative turn in planning theory and its implications for spatial strategy formation", *Environment and Planning B: Planning and Design*, 23: 217–34.

Jankowski, P., Nyerges, T.L., Smith, A., Moore, T.J. and Horvath, E. (1997) "Spatial group choice: a SDSS tool for collaborative spatial decision making", *International Journal of Geographical Information Systems*, 11(6): 577–602.

Jankowski, P. and Nyerges, T.L. (2001) "GIS-Supported Collaborative Decision Making: Results of an Experiment", *Annals of the Association of American Geographers*, in press March issue.

Jankowski, P. and Stasik, M. (1997) "Spatial understanding and decision support system: a prototype for public GIS", *Transactions in GIS*, 2(1): 73–84

Jarke, M. (1986) "Knowledge sharing and negotiation support in multiperson decision support systems", *Decision Support Systems*, 2: 93–102.

Jarvenpaa, S. (1989) "The effect of task demands and graphical format on information processing strategies", *Management Science*, 35(3): 285–303.

Jarvenpaa, S.L. and Dickson, G.W. (1988) "Graphics and managerial decision making: research based guidelines", *Communications of the ACM*, 31(6): 764–74.

Jessup, L. and Valacich, J. (eds) (1993) *Group Support Systems: New Perspectives*, New York, Macmillan Publishing Company.

Kahneman, D. (1992) "Reference points, anchors, norms, and mixed feelings", *Organizational Behavior and Human Decision Processes*, 51: 296–312.

Keeney, R. (1992) *Value-Focused Thinking*, Cambridge, Harvard University Press.

Kunreuther, H.C., Linnerooth, J. and others (1983) *Risk Analysis and Decision Processes*, Berlin, Springer-Verlag.

Kunreuther, H., Slovic, P. and MacGregor, D. (1996) "Risk perception and trust: challenges for facility siting", *Risk* 7(2): 99–108.

Lake, R.W. (ed.) (1987) *Resolving Locational Conflict*, New Brunswick, NJ, Rutgers Center for Urban Policy Research.

Lake, R.W. (1993) "Planning and applied geography: positivism, ethics, and geographic information systems", *Progress in Human Geography*, 17(3): 404–13.

Leitner, H., McMaster, R., Elwood, S., McMaster, S. and Sheppard, E. (1998) "Models for Making GIS Available to Community Organizations: Dimensions of Difference and Appropriateness", paper presented to NCGIA Specialist Meeting on Empowerment, Marginalization and GIS, Santa Barbara, CA.

Linnerooth-Bayer, J. and Löfstedt, R.E. (1996) "Introduction", *Risk* (7)2: 95–8.

Lober, D.J. (1995) "Resolving the siting impasse: modeling social and environmental locational criteria with a geographic information system", *American Planning Association (APA) Journal*, Autumn: 482–95.

McCartt, A. and Rohrbaugh, J. (1995) "Managerial openness to change and the introduction of GDSS: explaining initial success and failure in decision conferencing", *Organization Science*, 6: 569–84.

McGrath, J.E. (1984) *Groups: Interaction and Performance*, Englewood Cliffs, NJ, Prentice Hall.

McGrath, J.E. (1990) "Time matters in groups", in J. Galegher, R.E. Kraut and C. Egido (eds) *Intellectual Teamwork: Social and Technological Foundations of Cooperative Work*, Hillsdale, NJ, Lawrence Erlbaum Associates: 23–62.

McGrath, J.E. (1991) "Time, interaction, and performance (TIP): a theory of groups", *Small Group Research*, 22(2): 147–74.

Moore, T.J. (1997) "GIS, society, and decisions: a new direction with SUDSS in command", *Proceedings*, *AutoCarto 13*, v. 5, Seattle, Washington, American Society for Photogrammetry and Remote Sensing: 254–66.

National Center for Environmental Decision-Making Research (1998) http://www.ncedr.org/decision_path/

Nyerges, T.L. (1991) "Geographic information abstractions: conceptual clarity for geographic modeling", *Environment and Planning A*, 23: 1483–99.

Nyerges, T.L. (1993) "How do people use geographical information systems?", in D. Medyckyj-Scott and H. Hearnshaw (eds) *Human Factors for Geographical Information Systems*, New York, John Wiley & Sons, 37–50.

Nyerges, T.L. (1995) "Design considerations for group-based GIS: an example from transportation improvement program decision making", *Proceedings*, GIS for Transportation '95, Reno, NV, April 1–3, 1995: 261–82.

Nyerges, T.L. and Jankowski, P. (1997) "Enhanced adaptive structuration theory: a theory of GIS-supported collaborative decision making", *Geographical Systems*, 4(3): 225–57.

Nyerges, T.L., Moore, T.J., Montejano, R. and Compton, M. (1998) "Interaction coding systems for studying the use of groupware", *Journal of Human-Computer Interaction*, 13(2): 127–65.

Obermeyer, N. (1998) "The Evolution of Public Participation GIS", *Cartography and Geographic Information Systems*, 25(2): 65–6.

Obermeyer, N. and Pinto, J. (1994) *Managing Geographic Information Systems*, New York, Guilford.

Orlikowski, W.J. (1992) "The duality of technology: rethinking the concept of technology in organizations", *Organization Science*, 3(3): 398–427.

Ostrom, E. (1992) "The rudiments of a theory of the origin, survival and perform-ance of common property institutions", in D.W. Bromley (ed.) *Making the Commons Work: Theory, Practice and Policy*, San Francisco, ICS Press: 569–84.

Ostrom, E., Gardner, R. and Walker, J. (1994) *Rules, Games, and Common Pool Resources*, Ann Arbor, University of Michigan Press.

Pasquero, J. (1991) "Supranational collaboration: the canadian environmental experiment", *Journal of Applied Behavioral Science*, 27(1): 38–64.

Pestman, P. (1998) "Dutch infrastructure policies, public participation and the environment of the 1990's: the politics of interfering logics", in F.H.J.M. Coenen, D. Huitema and L.J. O'Toole (eds) *Participation and the Quality of Environ-mental Decision Making*, Dordrecht, Kluwer Publishers: 185–202.

Poole, M.S. (1985) "Tasks and interaction sequences: a theory of coherence in group decision-making interaction", in R.L. Street, Jr. and J.N. Cappella (eds) *Sequence and Pattern in Communicative Behaviour*, London, Edward Arnold: 206–24.

Poole, M.S. and Roth, J. (1989) "Decision Development in small groups IV: a typology of group decision paths", *Human Communication Research*, 15(3): 323–56.

Popper, F.J. (1981) "Siting LULUs", *Planning*, 47: 12.

Pretty, J. (1998) "Participatory learning in rural Africa: towards better decisions for agricultural development", in F.H.J.M. Coenen, D. Huitema and L.J. O'Toole (eds) *Participation and the Quality of Environmental Decision Making*, Dordrecht, Kluwer Publishers: 251–66.

Pickles, J. (ed.) (1995) *Ground Truth: The Social Implications of Geographic Information Systems*, New York, Guilford.

Pickles, J. (1997) "Tool or science? GIS, technoscience, and the theoretical turn", *Annals of the Association of American Geographers*, 87(2): 363–72.

Pinto, J.K. and Azad, B. (1994) "The role of organizational politics in GIS imple-mentation", *Journal of Urban and Regional Information Systems Association*, 6(2): 35–61.

Pressman, J.L. and Wildavsky, A.B. (1984) "Implementation: how great expecta-tions in Washington are dashed in Oakland: or, why it's amazing that federal programs work at all, this being a saga of the Economic Development Adminis-tration as told by two sympathetic observers who seek to build morals on a foun-dation of hopes" (third edn), Berkeley, University of California Press.

Raiffa, H. (1982) *The Art and Science of Negotiation*, Cambridge, Harvard Univer-sity Press.

Reagan, P. and Rohrbaugh, J. (1990) "Group decision process effectiveness: a competing values approach", *Group and Organization Studies*, 15: 20–43.

Rejeski, D. (1993) "GIS and risk: a three-culture problem", in M. Goodchild, B. Parks and L. Stayaert (eds), *Environmental Modeling with GIS*, New York, Oxford University Press: 318–31.

Renn, O., Webler, T., Rakel, H., Dienel, P. and Johnson, B. (1993) "Public participation in decision making: a three-step procedure", *Policy Sciences*, 26: 189–214.

Renn, O., Webler, T. and Wiedemann, P. (1995) *Fairness and Competence in Citizen Participation: Evaluating Models for Environmental Discourse*, Dordrecht, Kluwer Academic Publishers.

Rittel, H.W.J. and Webber, M.M. (1973) "Dilemmas in a General Theory of Planning", *Policy Sciences*, 4: 155–69.

Roberts, N.C. and Bradley, R.T. (1991) "Stakeholder collaboration and innovation: a study of public policy initiation at the state level", *Journal of Applied Behavioral Sciences*, 27(2): 209–27.

Rohrbaugh, J. (1989) "Demonstration experiments in field settings: assessing the process, not the outcome, of group decision support", *Harvard Business School Research Colloquium*, Boston, Harvard Business School, Publishing Division: 113–30.

Rosa, E.A. (1998) "Metatheoretical foundations of post-normal risk", *Journal of Risk Research* 1(1): 15–44.

Roschelle, J. and Teasley, S.D. (1995) "The construction of shared knowledge in collaborative problem solving", in C. O'Malley (ed.) *Computer Supported Collaborative Learning*, Berlin, Springer-Verlag: 69–97

Sackmann, S.A. (1991) "Uncovering culture in organizations", *Journal of Applied Behavioral Science*, 27(3): 295–317.

Sandman, P.M. (1993) *Responding to Community Outrage: Strategies for Effective Risk Communication*, Virginia, Fairfax, American Industrial Hygiene Association.

Sayer, A. (1984) *Method in Social Science*, London, Hutchinson Education.

Schank, G. (1998) "The extraordinary ordinary powers of abductive reasoning", *Theory and Psychology*, 8(6): 841–60.

Schneider, E., Oppermann, B. and Renn, O. (1998) "Implementing structured participation for regional level waste management planning", *Risk: Health, Safety and Environment*, 9(4): 379–95.

Sheppard, E. (1995) "GIS and society: towards a research agenda", *Cartography and Geographic Information Systems*, 22(1): 251–317.

Shiffer, M.J. (1998) "Multimedia GIS for planning support and public discourse", *Cartography and Geographic Information Systems*, 25(2): 89–94.

Simon, H. (1977) *The New Science of Management Decision*, 3rd edn, Englewood Cliffs, NJ, Prentice-Hall.

Simon, H. (1979) "Rational decision making in business organizations", *American Economic Review*, 69: 493–513.

Steinitz, C. (1990) "A framework for theory applicable to the education of landscape architects (and other design professionals)", *Landscape Journal*, 9(2): 136–43.

Stern, P.C. and Fineberg, H.V. (eds) (1996) *Understanding Risk: Informing Decisions in a Democratic Society*, Washington, DC, National Academy Press.

Susskind, L. and Cruikshank, J. (1987) *Breaking the Impasse: Consensual Approaches to Resolving Public Disputes*, New York, Basic Books.

Susskind, L. and Field, P. (1996) *Dealing with an Angry Public*, New York, The Free Press.

Susskind, L., Levy, P. and Thomas-Larmer, J. (2000) *Negotiating Environmental Agreements*, Washington, DC, Island Press.

Todd, P. (1995) "Process tracing methods in the decision sciences", in T. Nyerges, D. Mark, M. Egenhofer and R. Laurini (eds) *Cognitive Aspects of Human-computer Interaction for Geographic Information Systems*, Dordrecht, Kluwer, September 1995: 77–95.

Turk, A. (1993) "The relevance of human factors to geographical information systems", in D. Medyckyj-Scott and H. Hearnshaw (eds) *Human Factors in Geographical Information Systems*, New York, John Wiley: 15–36.

van der Schans, R. (1990) "The WDGM model: a formal systems view of GIS", *International Journal of Geographical Information Systems*, 4(3): 225–39.

von Winterfeldt, D. (1987) "Value tree analysis: an introduction and an application to offshore oil drilling", in P.R. Kleindorfer and H.C. Kunreuther (eds) *Insuring and Managing Hazardous Risks: From Seveso to Bhopal and Beyond*, Berlin, Springer-Verlag: 349–85.

Webler, T. (1995) "'Right' discourse in citizen participation: an evaluative yardstick", in O. Renn, T. Webler and P. Wiedemann, *Fairness and Competence in Citizen Participation*, Boston, Kluwer: 35–86.

Wood, D.J. and Gray, B. (1991) "Toward a comprehensive theory of collaboration", *Journal of Applied Behavioral Sciences*, 27(2): 139–62.

Zadeh, L. (1965) "Fuzzy sets", *Information and Control*, 8: 338–53.

# 3 Methods and tools for participatory, spatial decision support

## Abstract

In Chapter 2 we introduced a general framework for understanding collaborative spatial decision making. The framework helps lay out a systematic way of conducting a task analysis as the basis of a user needs assessment for setting up collaborative decision making approaches to problem solving and decision making and for analysing collaborative decision making processes. Guided by that framework, we now present specific methods and tools for participatory spatial decision support, and the hardware and software architectures to implement decision support. Methodologies and tools for participatory group decision making come from many sources. They include work on GIS extensions aimed at improving its decision support capabilities, work on group support systems technology as well as theoretical and empirical studies of its use. Other sources include work on capturing the dynamics of argumentation, research on the human dimensions of groupware and computer networking, and critiques of GIS as a construction of positivist thinking, constraining alternative views of reality that otherwise might broaden the decision making discourse. These sources bring various viewpoints of decision making that can be generalized as a decision analytical and collaborative approach. The analytical approach uses mathematical models to analyze structured parts of a decision problem, leaving the unstructured parts for the decision makers' judgement. The collaborative approach views decision making as an evolutionary process that progresses from an unstructured discourse to a problem resolution using discussion, argumentation, and voting. We argue that both approaches are needed in a group decision support environment and that in order to support effectively group participation in decision making, collaboration and decision analysis tools must be integrated. We present a variety of methods and tools for participatory group decision support.

Methodologies and tools for participatory group decision making come from many sources. They include work on GIS extensions aimed at improving its decision support capabilities (Densham 1991), work on group support systems technology as well as theoretical and empirical studies of its use (Jessup and Valacich 1993), work on capturing the dynamics of argumentation (Conklin and Begeman 1989), research on the human dimensions of groupware and computer networking (Oravec 1996), and critiques of GIS as a construction of positivist thinking, constraining alternative views of reality that otherwise might broaden the decision making discourse (Lake 1993, Sheppard 1995, Pickles 1996). These sources bring various viewpoints on decision making that can be generalized as a decision analytical and collaborative approach. The analytical approach uses mathematical models to analyze structured parts of a decision problem, leaving the unstructured parts for the decision makers' judgement. The collaborative approach views decision making as an evolutionary process that progresses from an unstructured discourse to a problem resolution using discussion, argumentation, and voting. We argue that both approaches are needed in a group decision support environment and that in order to support effectively group participation in decision making, collaboration and decision analysis tools must be integrated (Bhargarva, Krishnan and Whinston 1994, Stern and Fineberg 1996).

In Chapter 2 we introduced a general framework for understanding collaborative spatial decision making. The framework helps lay out a systematic way of conducting a task analysis (needs assessment) for setting up collaborative decision making approaches to problem solving and decision making, and for analysing collaborative decision making processes. Guided by that framework, we now present specific methods and tools for participatory spatial decision support and the hardware and software architectures to implement the support.

## 3.1 Macro-micro decision strategy for methods and tools

The reader may remember from section 2.2 that a macro decision strategy contains many decision tasks, each task being a phase in the overall process of decision making and problem solving. Each of the macro-phases can be characterized in terms of micro activities. The matrix in Table 3.1 represents an expanded view of Table 2.1 as a decision process flowing from a normative model of decision making. The decision strategy consists of three macro-phases: intelligence on criteria, design of options set, and choice of options (Simon 1979, Renn *et al.* 1993); each phase is composed of four micro activities: gather, organize, select, and review (Bhargarva, Krishnan and Whinston 1994, Simon 1977, Dewey 1933). It is possible to use this table to relate types of decision support methods and tools to the phase-activities. Each cell in the matrix contains a specific method/tool, which we identified as potentially useful for supporting a specific phase-activity. We have found this type of cell by cell elucidation of a

*Table 3.1* Methods and tools for CSDM derived from macro-micro decision strategy

| *Micro-decision strategy activities* | Macro-decision strategy phases | | |
|---|---|---|---|
| | *1. Intelligence about values, objectives and criteria* | *2. Design of a feasible option set* | *3. Choice about decision options* |
| **A. Gather ...** | participant input on values, goals and objectives using **information management** and **structured-group process techniques** | **data and models** (GIS and spatial analysis, process models, optimization, simulation) to generate options | values, criteria and feasible decision options using **group collaboration support methods** |
| **B. Organize ...** | goals and objectives using **representation aids** | an approach to decision option generation using **structured-group process techniques** and **models** | values, criteria and feasible decision options using **choice models** |
| **C. Select ...** | criteria to be used in decision process using **group collaboration support methods** | decision options from outcomes generated by **group process techniques** and models | goal- and consensus-achieving decision options using **choice models** |
| **D. Review ...** | criteria, resources, constraints, and standards using **group collaboration support methods** | decision options and identify feasible options using **information management** and **choice models** | recommendation(s) of decision options using **judgement refinement techniques** |

"decision support agenda" useful for unpacking the complexity of a variety of macro-micro decision strategies — each strategy having an appropriate support for phase-activity as needed by the situation at hand.

Some methods/tools address phase 1, as for example with *representation aids* that organize discussions and/or present information about goal(s) and objective(s). Other methods/tools might help with structuring a problem in phase 2 in the sense of designing an approach for analysis and generating options. This is where GIS plays a significant role, starting out with the computations necessary to manipulate spatial characteristics of decision options. Then, in phase 3 decision analysis methods help with evaluating options, as in the case of *choice models* and *judgement refinement techniques*. Consequently, a suite of methods/tools is likely needed to address complex decision problems, since no single system or environment addresses all phases (at least at the time of writing in the spring of 2000).

In the intelligence phase of the decision process the activity phases 1A–1D are meant to encourage the articulation of values, objectives, criteria, constraints and standards that impinge on opportunities for solutions. The methods and tools identified as potentially useful for these activity phases are:

1A **Information management** and **structured-group process techniques** to help gather participant input on values underlying the decision process and resulting from them goals and objectives;
1B **Representation aids** to help organize goal and objectives into structures that can be analyzed, using symbols, text, and graphics;
1C **Group collaboration support methods** to help select criteria on the bases of objectives, and articulate those criteria in terms of how they will be measured;
1D **Group collaboration support methods** to aid in the review of criteria, constraints, and standards for option generation.

The intelligence phase can be supported by open group discussion, but more commonly stakeholder interviews have been performed to gather information about values. The goal is to draw ideas out and synthesize them for each stakeholder group. In organizing the material from stakeholder interviews, **information management techniques** can be useful. An example of one such technique that has shown to be useful is "value tree analysis" (von Winterfeldt 1987).

Value trees are hierarchical (tree-like) representations of values, objectives and criteria, where values are roots, objectives are nodes to branches and criteria are leaves at the ends of branches. Objectives stem from values as concerns that are to be considered. Criteria, stemming from objectives, are measurable characteristics used to evaluate performance of options. The "tree" is a way of organizing ideas/issues into a hierarchy of values, objectives and criteria. Once the values, objectives and criteria have been identified for each stakeholder group and represented as a "value tree", then it is recommended that the individual trees be consolidated into an overall tree so that all groups know where they "stand" in regards to overall concerns. Some roots, branches (nodes) might be shared and others might not.

If the process of eliciting values, objectives and criteria were to take place in a collaborative setting, rather than through stakeholder interviews, then **structured-group process techniques** might be useful. One such structured-group process technique is the technology of participation (ToP), developed and used by the Institute for Cultural Affairs (Spencer 1989).

ToP helps people share an understanding of what is at issue by attempting to generate consensus during strategic planning. The process has four steps:

1   generate and cluster ideas from everyone;
2   identify constraints and barriers that might impinge on those issues;
3   prioritize issues in line with the constraints; and
4   flesh out a plan developed from the prioritized issues.

At least the first two steps are relevant to this phase of decision making, with the third and fourth steps being relevant to phases 2 and 3, respectively. In step 1 a facilitator would help the group elicit, cluster and label the values, objectives and criteria for a decision situation. In step 2 a facilitator would ask the group to articulate the resources, constraints (barriers) and standards that would influence successful outcomes of the situation.

Value trees can be constructed off-line by processing stakeholder interviews, or they may be constructed on-line with support of a structured-group process technique. To the authors' knowledge, tools to implement this particular technique consist currently of only paper and pencil or drawing package. Alternatively, one might use a software package such as Concept Systems (Trochim 1989) to assist with the synthesis. The Concept Systems package has been designed explicitly for group use. Such a package helps articulate and cluster ideas using "concept maps" to show how people can construct a map (representation) of their "ideas" in a concept space (as opposed to a geographic map) for all ideas presented.

In the review activity, a decision is made as to whether a single, consensus-based set of criteria will be used or multiple (stakeholder) sets of criteria will be used in the design phase involving options generation. If a single set, then which set, or a combined set? This review might involve negotiation among the groups to see which criteria, hence objectives, hence values, move forward to the next phase. In a sense this is actually negotiation of a problem definition, as certain criteria (called attributes in a GIS context) will describe options in a different way from other criteria (attributes). This negotiation leads to specifying types of options that would be considered as feasible solutions to the problem and can be aided by **group collaboration support methods**.

In the design phase (phase 2) it is appropriate to allow each group to generate the options they feel can address the goal of the decision problem in line with the criteria that have been established. To do this, groups would:

2A  gather **data** about the outcomes of criteria to be used as a basis for option generation, i.e. criteria, which drive the generation of scenarios that become decision options. Different types of **models** can also generate the data;

2B  organize and apply an approach for generating feasible options using one or more **structured-group process techniques** or computer **models**. Structured-group process techniques may include brainstorming, Delphi, or technology of participation. Models may range from suitability models implemented in GIS to optimization models generating decision option scenarios that satisfy the decision objectives;

2C select the full array of options to be considered, despite immediate constraints, resources, and standards identified in activity-phase 1D;

2D review the option set(s) in line with constraints, resources and standards from activity phase 1D and select feasible decision options. The selection of feasible options can be supported by **information management** tools and **choice models**.

In the design phase, groups work in activity-phase 2A to identify the basis for creating options. That is accomplished by first identifying primary attribute(s) that allow differentiation among fundamental choices, and then identifying secondary attributes that are used in determining "feasibility". For example, a "land" parcel is a primary attribute that, once set, allows a decision analyst to examine the entire range of land parcels for feasible options. In activity phase 2B various techniques can be used to process the secondary attributes, e.g. parcel size, cost per hectare, tax rate, to establish a feasibility (minimum threshold) for each option to be included in the feasible set. Examples of these techniques include decision option enumeration through group conversation, standards articulation and minimum thresholds, GIS analysis, and/or modeling such as location–allocation optimization analysis. No single tool has surfaced that specifies the "best way" to identify options, but GIS is one such tool that has grown in popular use because it allows large data sets to be processed to find feasible options. The use of a technique must properly match the character of a decision situation to be effective. For example, a location–allocation model can be used to identify "optimal" options under varying conditions of an objective function. Those options might be the "best cases" to be used in starting a search for decision options, relaxing the objective function to one of "satisficing" rather than "optimizing". Thus, the challenge of activity-phase 2B is to have participants recognize which kind of problem they are facing, in addition to what data are available, in order to provide the most effective decision support for option generation.

The activity phase 2C has participants listing the full array of feasible options. This list might come from various stakeholder groups, or it might have been generated as the result of considering different parameters and constraints in models describing a decision situation. During activity-phase 2D participants from different groups can come together to share an understanding of the different options that each sees as feasible options for the problem. The activity is one of reviewing the principal goal of the process, understanding that certain equitable concerns might need to be addressed. The equitable concerns might in fact be "geographic area" related, but will probably be idiosyncratic to the decision situation at hand. Because of these differences it is possible to conceive of various scenarios that might be generated, each scenario having an option set associated with it. The result of phase 2 could be a plan describing one or more scenarios each with option set(s) for consideration in phase 3.

The choice phase (phase 3) involves evaluation of the option set(s), within one or more scenarios. In general the evaluation process involves the following phase-activities.

3A Gather information using **group collaboration support methods** about how to proceed with evaluation of decision options based on values and criteria articulated in phase 1 and options identified in phase 2.

3B Organize an approach to evaluating decision options using **choice models**.

3C Select a prioritized/ordered list of options, using choice models, consensus building, negotiation, mediation, or arbitration.

3D Review how the recommendation(s) address decision situation goal(s) in line with values consideration, using **judgement refinement techniques**.

In activity-phase 3A, decision participants gather information from phase 1 in terms of values and criteria to be considered, and from phase 2 in terms of option set(s) for one or more scenarios. At this time it is possible that various stakeholder groups might elect to enter into discussion about what values, objectives, criteria, resources, constraints and option lists have been identified to this point, before evaluation (in many cases a negotiated evaluation) takes place. At this time the comparison of individual values and decision criteria could be performed, showing how each stakeholder compares to the overall group. A value and criteria assessment is a good lead into the prioritization of options that takes place within a group. Next, in activity phase 3B participants might use GIS and/or multiple criteria choice models. Standard GIS operations allow only a non-compensatory evaluation of decision options to be performed. That is, a threshold range(s) is set for one or more criteria, and all options that fall within that range pass through the "filter". With choice models, a compensatory (tradeoff) analysis can be performed. In compensatory analysis, when evaluating a decision option, a satisfactory outcome for a high priority criterion can "compensate" for less than satisfactory outcomes on lower priority criteria. For example, a less than desirable size of a given land parcel can be compensated by a favorable tax rate.

In activity-phase 3C "selecting a choice" can be accomplished through option evaluation using choice models, consensus voting, negotiation, mediation and/or arbitration. Consensus voting can be performed by combining the prioritized lists of all separate stakeholder groups. Such a consensus voting might be the start of a negotiation process. Negotiation might involve taking the overall list of priorities as a start and then discussing how one site might move above another if certain of the criteria were weighted in a different way. Option sets might be devised to help with this negotiation. Such sets could be developed as an "equity agreement" whereby certain options from a given area are discussed in relation to each other, but the options in different sets are not discussed in relation to each other. That is, "if you get one, then I get one, and so does she". Alternatively, the participants might turn to mediation or arbitration to effect a decision. When negotiation does not work, mediation using a third party go-between might prove beneficial. The third party helps to balance the negative feelings among the (commonly) two perspectives (sides). Having more perspectives complicates the mediation process even further. Discussing the differences using "decision aids" can help externalize the disagreements. In yet more extreme conflict circumstances, arbitration might be needed.

Arbitration is in a sense a mandated mediation, i.e. it is usually required by a court order for the good of all sides concerned. Legal interpretations are generally needed to clarify what is really at stake, and what might happen if a settlement is not reached, e.g. a fine levied. If all of these processes fail in an attempt to reach a recommendation, then the court usually takes over as a "last resort" effort to establish a decision outcome. Regardless of what party establishes the recommended decision outcome, a review takes place. Such a review can be supported by judgement refinement techniques. During the review of recommended choice(s) in activity-phase 3D, the options are discussed in terms of the original values devised from phase 1. Mentally returning to original goals can be the final "comfort check" on whether the entire process played out in an equitable way. Writing up the (final) report to document what is to be recommended for implementation is the final activity in the decision process. As the next step is implementation to take action, the decision process followed by implementation and monitoring are all three together considered the macro-phases of a problem solving process.

The above decision process is but one of many normative processes that can be established as an agenda agreed upon by participants — or at least the convenor and those responsible for the process. A different set of macro-phases might shorten or lengthen the process agenda in different situations. In any case, no matter how many macro-phase steps might exist, the micro activities for each phase set the stage for requirements analysis.

## 3.2 System requirement analysis for collaborative spatial decision support

System requirements for computerized support of collaborative spatial decision making depend on the 25 aspects of EAST2 as the basis of a comprehensive needs assessment presented in section 2.2. Among the more important aspects of those needs for decision support are understanding the type of participants, e.g. novices or experts, and meeting arrangements for collaborative work, e.g. face-to-face meeting, long-distance conference, different place/different time group work. There are, however, sufficiently common tasks such that computerized decision support tools can be developed to support a range of participants on the various phase-activities in the decision strategy in the context of various meeting venues. Meeting participants could collaborate on design and construction of various geographic alternatives, sharing interactive mapping tools over a local area network (Faber, Wallace and Cuthbertson 1995). The evaluation of collaboratively designed alternatives can be carried out with multiple criteria evaluation techniques enhanced by voting tools (Malczewski 1996). The evaluation results can be visualized on special-purpose maps capable of geographically representing consensus solutions (Jankowski and Stasik 1997, Armstrong and Densham 1995).

Based on the knowledge abilities of decision participants (as they range across experts to novices in using spatial decision support tools) and meeting venues (as they range across place and time), the following design requirements are common.

1. A spatial decision support system for collaborative work should offer decisional guidance to users in the form of an agenda (such as that presented in the 12 phase-activities in Table 3.1).
2. A system should not be restrictive, allowing the users to select tools and procedures in any order.
3. A system should be comprehensive within the realm of spatial decision problems, and thus offer a number of decision space exploration tools and evaluation techniques.
4. The user interface should be both process-oriented and data-oriented to allow an equally easy access to task-solving techniques, as well as maps and data visualization tools.
5. A system should be capable of supporting facilitated meetings and, hence, allow for the information exchange to proceed among group members, and between group members and the facilitator. It should also support space- and time-distributed collaborative work by facilitating information exchange, electronic submission of solution options, and voting through the internet.
6. A system functionality should include extensive multiple criteria evaluation capabilities, sensitivity analysis, specialized maps to support the enumeration of preferences and comparison of alternative performance, voting, and consensus building tools.
7. A system should provide necessary functionality to support needs of an advanced user without overwhelming a novice who needs a user-guiding interface.

In the next three subsections we present:

a. various meeting venues as "decision settings";

b   certain hardware configurations that are perhaps more useful than others within the respective settings; and

c   the functionality of a wide array of software tools and their under-lying decision support methods.

### 3.2.1 GIS-supported participatory decision making in various meeting arrangements

Four types of meeting arrangements are feasible, each described in terms of place and time, which tend to constrain and/or foster human-computer-human interaction. Several advantages and disadvantages of various meeting arrangements can be described (see Table 3.2).

The most common type of meeting arrangement ("conventional meeting") supports a group convened in the same place at the same time, whereby collaborative work can be performed in a conference room with a single computer or multiple computers connected on a local area network. A second type of arrangement ("storyboard meeting") involves the same place but different time, whereby collaborative work is performed by using leave behind materials, supported by a local, or wide area computer network. A third type of meeting arrangement ("computer conferencing

*Table 3.2* Meeting arrangements as a combination of time and place

|  | Same time | Different time |
|---|---|---|
| **Same place** | **Conventional meeting** | **Storyboard meeting** |
|  | *Advantage:* | *Advantage:* |
|  | – face-to-face expressions | – scheduling is easy |
|  | – immediate response | – respond anytime |
|  |  | – leave-behind note |
|  | *Disadvantage:* | *Disadvantage:* |
|  | – scheduling is difficult | – meeting takes longer |
|  |  | – difficult to maintain in the long run |
| **Different place** | **Conference call meeting** | **Distributed meeting** |
|  | *Advantage:* | *Advantage:* |
|  | – no need to travel | – scheduling is convenient |
|  | – immediate response | – no need to travel |
|  |  | – submit response anytime |
|  | *Disadvantage:* | *Disadvantage:* |
|  | – limited personal perspective from participants | – meeting takes longer |
|  | – meeting protocols are difficult to interpret, difficult to maintain meeting dynamics | – meeting dynamics are different from normal meeting ("netiquette" instead of face-to-face etiquette) |

meeting", as in a teleconferencing meeting) involves different places but the same time, whereby collaborative work can be performed using interactive desktop audio and video, supported by a wide area network, a dedicated wide band-width telephone line, or a satellite link. A fourth type of meeting arrangement ("distributed meeting") involves different places and different times, whereby collaborative work can be performed using e-mail, broadband network, and network-resident multimedia tools. Each of the four types of meeting arrangements and the interaction that each encourages can be best supported with different hardware configuration requirements.

Two examples of decision rooms for same place–same time collaboration exist in the Department of Geography at the University of Washington, informally called medium technology and high technology group environments. The medium technology environment supports 30 students working at six tables, one computer per table. The high technology environment supports 18 computers, one computer for each person.

The medium technology environment consists of a local area network with seven Pentium III PCs (see Figure 3.1). Each of six of the Pentium PCs have 24-inch monitors on tables with six chairs each. The six person tables were chosen to implement work (stakeholder) groups, i.e. a space convenient for small group discussion. A keyboard cord and mouse cable are long enough at each table such that they can be used across the table depending on who chooses to operate the computer. The facilitator/chauffeur workstation, which has a 20-inch monitor, is in the middle of the room. The middle of the room was chosen to focus attention when "group as a whole" displays are shown. A $1024 \times 768$-monitor resolution, RGB projector is attached to the chauffeur workstation and projects to a wall-mounted screen. The facilitator can roam the room, and easily point to various displays using an infrared pointer.

The high technology environment includes 18 Pentium III computers with 17-inch monitors connected to the same local area network (LAN) as mentioned above. However, the computers are located in a different room than those mentioned above. The computers are arranged in a small U-shaped layout (seven computers) inside a larger U-shape (12 computers). One station in the smaller "U" is specially configured for use by the facilitator/chauffeur. That machine is at the end of the "U" and has attached to it a $1024 \times 768$-monitor resolution RGB projector to drive the public display screen (see Figure 3.2).

### 3.2.2 Hardware requirements for meeting arrangements

Various hardware configurations can be used to support the four types of meeting arrangements. Generally, three computer hardware configurations are possible (see Table 3.3). They are listed in order of increasing costs and increasing flexibility. Configuration 1 is a standalone personal

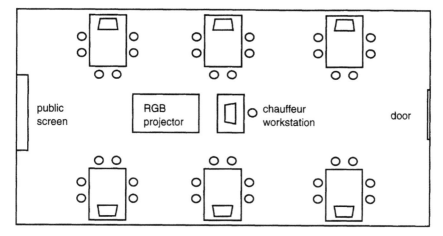

*Figure 3.1* Medium technology, face-to-face, collaborative, decision room

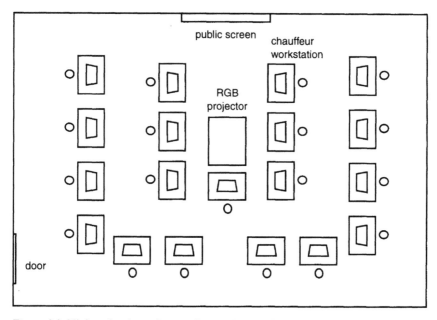

*Figure 3.2* High technology, face-to-face, collaborative, decision room

computer (PC), a laptop environment, or a network appliance (such as a "dumb" computer terminal). Configuration 2 contains a set of distributed PCs connected on a low-band network (low-band denotes here the capability to transmit text, audio, and still video). Configuration 3 contains a set of distributed PCs connected on a high-band network (high-band denotes here the capability to transmit real-time video). Collaboration for each of the meeting arrangements can be supported as follows.

**Same time–same place** (conventional face-to-face meeting) collaboration can be supported by any of configurations 1, 2 or 3. However, a network server is required if multiple parties are to participate simultaneously, hence configurations 2 and 3 are better than configuration 1.

**Different time–same place** (storyboard meeting) collaboration can be supported by any computer configuration. All configurations can be equally beneficial as long as "place" is interpreted to mean the same building (floor). In this meeting arrangement, because time is a relaxed constraint, participants can go to a common, stand-alone computer, perform work, and "leave behind" on the hard disk, the results of their decision activity.

**Same time–different place** (conference-call meeting) collaboration can be supported by configurations 2 and 3, but the different place for configuration 2 is limited to audio and still-motion video conferencing. This type of collaboration can be characterized by the highest network bandwidth consumption. A more desirable is configuration 3 as it supports the interaction experience of same place–same time meeting environment.

**Different time–different place** (distributed meeting) collaboration can be supported by configurations 2 and 3. Since this is not a meeting-like participatory interaction the configuration 2 can be as effective as the configuration 3.

### 3.2.3 Software requirements

In section 3.1 we described a macro-micro decision strategy consisting of phase-activities used in collaborative spatial decision making. Each of the phase-activities can be supported with various decision aid methods and tools, as outlined in Table 3.1. Methods and tools have been synthesized from a combination of literature on management decision support systems, spatial decision support systems, group support systems, and the

*Table 3.3* Meeting arrangements supported by computer hardware configurations

| Meeting arrangement | Configuration | | |
|---|---|---|---|
| | *Configuration 1: stand-alone PC* | *Configuration 2: low-band network* | *Configuration 3: high-band network* |
| same time–same place | fair | good | good |
| different time–same place | good | good | good |
| same time–different place | – | fair | good |
| different time–different place | – | good | good |

young literature on collaborative GIS. These methods and tools from Table 3.1 are classified at three levels of functional capabilities (see Table 3.4). Some or all of the decision aiding methods and tools could potentially be useful for a GIS collaborative activity, but of course this depends on the situation. The methods and tools are listed in order of most basic to most sophisticated, and treated to be building blocks of a system. Level 1 capabilities are likely to be used the most and, therefore, appear in software packages as basic support since they satisfy a basic cognitive need for information manipulation. In contrast, capabilities in level 2 are specialized enough to be needed only by some groups who have a requirement for data analysis. Level 3 capabilities are the most sophisticated, and are more complicated than those at other levels when it comes to use. Overall, the decision methods and tools, which have been developed as stand-alone

*Table 3.4* Decision aiding methods and techniques for collaborative spatial decision support (adapted from Nyerges *et al.* 1998)

---

*Level 1: basic information handling support*
(a) Information management: storage, retrieval and organization of spatial data and information (e.g. distributed database management system support)
(b) Visual aids: manipulation (analysis) and expression (visualization) techniques for a specific part of decision problem (e.g. shared displays of charts, tables, maps, diagrams, matrix and/or other representational formats)
(c) Group collaboration support: techniques for idea generation, collection, and compilation; includes anonymous input of ideas, pooling and display of textual ideas, and search facilities to identify possible common ideas (e.g. data and voice transmission, electronic voting, electronic white boards, computer conferencing, and large-screen displays)

---

*Level 2: decision analysis support*
(d) Option modeling: methods of generating decision options. These include a variety of computational models from static spatial location models (e.g. suitability analysis in GIS) through optimization models (e.g. location–allocation models) to dynamic models that predict the behavior of real-world processes (e.g. hydrological models of river flow, or water pollution contribution based on effluent release)
(e) Choice models: integration of individual criteria across optional choices (e.g. multiple criteria decision models using multiple attributes and multiple alternatives for systematically weighted rankings or preferences)
(f) Structured-group process techniques: methods for facilitating and structuring decision making (e.g. brainstorming, Delphi, modified Delphi and technology of participation)

---

*Level 3: group reasoning support*
(g) Judgement refinement/amplification techniques: quantification of heuristic judgement processes (e.g. sensitivity/trade-off analysis for comparing project options, Bayesian analysis, or social judgement analysis for tracking each member's judgements for feedback to the individual or group)
(h) Analytical reasoning methods: perform problem-specific reasoning based on a representation of the decision problem (e.g. using mathematical programming or expert systems guided by automatic mediation, parliamentary procedure, or Robert's Rules of Order, identifying patterns in a reasoning process)

---

capabilities in the past, are now being brought together in an integrated manner for enhanced user benefit.

In addition to the integration of capabilities, it is also important to point out that the meeting arrangement can have a tremendous influence on the nature and use of the capability. Many of the capabilities were originally conceived for face-to-face interaction in a same-time and same-place decision activity. However, relaxing those assumptions about time and place could have a major influence on the usefulness of the methods and tools. Implementation of such tools in a distributed-meeting, i.e. different time and different place setting increases the challenge for system development. A reduction in face-to-face interaction is likely to require more support from software to assist in managing the distributed interaction.

Many of the functional capabilities presented in Table 3.4 can be useful to various kinds of decision tasks in various kinds of decision situations. Determining which capabilities to use is a matter of "fit" with a situation. To help with fostering an understanding of where a fit may exist, it is possible to frame the issue in terms of a system requirements analysis. A requirement analysis differs from a needs assessment in the sense that system requirements address and implement solutions for information needs. We can address the requirements analysis in a systematic manner as laid out in Table 3.5. Here we use the Renn *et al.* (1993) three-phase macro strategy down the left side of the table, and populate each of the phases with the micro strategy phase-activities (tasks) of Table 3.1. Filling in the cells is a matter of identifying which functional capabilities from Table 3.4 can (potentially) address the tasks of a particular decision situation as we did in the discussion of Table 3.1. We leave that exercise up to the reader when considering a specific situation.

Implementations of some of the capabilities listed in Table 3.4 have not yet found their way into commercial software products in a systematic way. In the discussion that follows in the next section, we focus more on capabilities that have found their way into commercially available software — describing methods and tools identified over the past several years.

## 3.3 Decision support capabilities

Despite the "orderly" categorization of the capabilities in Table 3.4, the advances that have been made in recent years point out that "cross-fertilization" of the capabilities is really what is important. Thus, in the discussion that follows we will see this cross-fertilization in terms of some capabilities being discussed in the "use context" of others.

### Level 1: Basic information handling support

*(a) Information management*   Data and information management is one of the three core geographic information technologies due in most part to

*Table 3.5* Requirements analysis structure for collaborative decision support

| Phase-activities | Decision aiding techniques (labels from Table 3.4) | | | | | | | |
|---|---|---|---|---|---|---|---|---|
| | *a* | *b* | *c* | *d* | *e* | *f* | *g* | *h* |
| 1. Intelligence | | | | | | | | |
| A. Gather | | | | | | | | |
| B. Organize | | | | | | | | |
| C. Select | | | | | | | | |
| D. Reflect | | | | | | | | |
| 2. Design | | | | | | | | |
| A. Gather | | | | | | | | |
| B. Organize | | | | | | | | |
| C. Select | | | | | | | | |
| D. Reflect | | | | | | | | |
| 3. Choice | | | | | | | | |
| A. Gather | | | | | | | | |
| B. Organize | | | | | | | | |
| C. Select | | | | | | | | |
| D. Reflect | | | | | | | | |

large amounts of data being processed. Information management for groups can be supported by virtual integration strategies, which rely on network retrieval of copies of data in a transparent fashion. In addition, distributed data management systems, e.g. Oracle and MS SQLServer, are two major competitors in the market that support attribute data retrieval and storage across local and wide-area computer networks. Although all GIS have basic data management capabilities, few GIS packages have distributed data management capabilities built into the spatial data manager component of the system. An exception is the extension to Oracle 8 for managing spatial data. Oracle8i Spatial can store and manage spatial data directly. Distributed data management is a basic requirement for group support, but decision makers would seldom use this technology directly, because summary information is provided only with customization.

However, even before summary information can be provided, that summary information is commonly based on fundamental categories of data that are basic to spatial data processing. For that reason, group-based database design languages are included in this category of decision support — although such languages might be used more aptly by decision analysts rather than decision makers themselves. Nonetheless, database designs drive the type of analysis performed for option generation. It makes sense to have a group review of the basic data categories plus attributes, hence criteria that will be used to frame a decision problem. Value tree analysis can be used to specify a database design, since a value tree maintains a relationship between values, objectives, and criteria for decision problems.

Although languages to support database design have tended to be rather arcane and numerous, there is a growing standard building in what is called the Unified Modeling Language (UML) (Rumbaugh, Jacobson and Booch 1999). Some GIS vendors, e.g. ESRI, have picked up on the need to make UML a viable way of specifying systems design and include a module in Arc/Info 8.0 to create such specifications. In addition, a recently published guide to geospatial database design called "Modeling Our World" (Zeiler 1999), contains a section on the use of UML that will help promote standards in database design.

Use of a group-based GIS has a requirement not only for access to shared data, but applications as well. Microsoft Windows COM (Common Object Model) and COM+ technology (released with Windows 2000), and Sun MicroSystems Enterprise Java Beans technology are two popular approaches to object linking and embedding. These are breakthroughs in application development, as we move from applications-centered data processing to document-centered data processing, as well as extending interactive data processing to the internet. Both of these developments foster integrated applications support for data management, and can only help with integration of visual representation and analysis techniques — the other two core technologies in GIS.

*(b) Visual aids*   Computer mapping display technology has been a mainstay in GIS as one of the three core technologies. Small platform (increased performance) GIS packages are at the stage where GIS can be on every desktop. All the major vendors are claiming at least 25,000 copies distributed. User interface development to support application-specific representations is a major growth area across many market sectors, as software is being customized to create representations suitable for specific work activities. In addition, multimedia support is appearing in GIS packages, as indicated by photo and sound manipulation capabilities in small platform packages. Charts, diagrams and tables may be linked to those representations to enhance information presentations. For example, Geo-ChoicePerspectives software for collaborative spatial decision support (see section 3.6.3) makes use of "histobar" displays (see Plate 1). It is possible to set the histobar interpretation such that high bars (lots of colorful graphic) indicate more preferred habitat development sites. Sometimes low data values in a database, e.g. for cost of redevelopment, should be shown as high (more preferred) bar displays. For other attributes, such as site size, a mid-range of data values, e.g. between 3–5 acres, is preferred for (re)development of a habitat area.

One challenge for graphical representation involves moving toward distributed display capabilities, following on the heels of distributed data management, and leading to collaborative document support. Geo-ChoicePerspectives implements shared displays in a local area network using map view file transfers. A more convenient approach for users is a "shared whiteboard" solution, as in GroupSystems Corporations' Group Systems for Windows, one of several group support systems (GSS) on the market. The shared whiteboard capability can support group notations on maps, providing for a "group writing space" as a storyboard where notations are left in a common space for others to comment on.

Another challenge for graphical representation involves integrating maps with other decision support methods and tools (e.g. models) using highly interactive, exploratory map displays. An example of such an interactive map called *value path/map* (see Plate 2) comes from a prototype called DECADE (**D**ynamic, **E**xploratory **Ca**rtography for **De**cision Support) (Jankowski, Andrienko and Andrienko 2001).

The value path/map in Plate 2 presents a single map display dynamically linked with the parallel coordinate plot. When the user points with the mouse on some option represented in the map, the respective object is highlighted in the map, and the corresponding value path is highlighted in the parallel coordinate plot. Alternatively, the user may point at some line segment in the plot, and the whole value path becomes highlighted, as well as the position of the corresponding option in the map. Such value path/map integration makes it easy to evaluate any option with regard to multiple decision criteria. Additionally, the user may "fix" highlighting by clicking on an object in the map or on a line in the plot. The selected object remains highlighted when the mouse cursor moves out of the display or points at other objects. This enables the comparison of value paths of two or more decision options.

*(c) Group collaboration support*   Group collaboration techniques are at the heart of GSS. These techniques support basic communication. Several GSS, including their capabilities, are reviewed in Bostrom, Watson and Kinney (1992). The capabilities make use of hardware technology that includes using data and voice transmission, electronic voting, electronic white boards, computer conferencing, and large-screen displays. Electronic mail (e-mail) is a crude form of communication through which one can broadcast a message to all members of a group. In a GSS, participants are supported with a structured conversation environment to carry on threaded conversations. The structure in large part comes from being able to design a meeting agenda with topics customized to the discussion at hand. The meeting agenda has topics supported by operations such as idea generation, collection, display, discussion, and voting. All of these operations lead toward support of consensus building. Simultaneous and anonymous idea generation from any computer linked in a group is one of the

major efficiencies brought about by GSS tools. File folders as "collection bins" for sorting through the ideas and then placing them in categories assists in synthesizing ideas. The synthesis by the group leader/facilitator can then be redisplayed to all participants in the group. Voting on lists of ideas, when done iteratively to sort through agreement, is a method for developing consensus among participants.

Implementation of GSS capabilities with GIS depends to a large degree on the operating system environment, and particularly the user interface support tools. Integration can range from loose to tight coupling (Nyerges 1993). One way to implement such capabilities in a GIS environment is to do it alongside GIS, relying on a similar operating system and user interface such as Microsoft Windows. This type of strategy was used in two GIS courses at the University of Washington to implement a collaborative learning and critical thinking support capability, using GroupSystems for Windows (GroupSystems 2000) and ArcView 2.1 (ESRI 1995), but was thought to be too cumbersome to be used by student groups undertaking GIS projects (Nyerges and Chrisman, 1994). Another approach is to rely on the GSS environment to provide hooks to GIS functions; these functions are invoked from a menu with the GSS user interface (Faber, Wallace and Miller 1996). However, such approaches limit the usefulness of both packages, as low-level data passing does not exist to provide a document-centered approach to information use. For this reason, development has been performed to make use of the dynamic data exchange functions within Microsoft Windows 95 to provide a direct connection for pooling votes from decision maker workstations. Such an approach was used to develop a loose-medium coupling between the ArcView GIS 3.2 and multiple criteria evaluation software as implemented in GeoChoicePerspectives.

Electronic voting belongs to the core of group collaboration support. Several types of voting are possible in group decision support. Traditional non-ranked voting, i.e. one person, one vote, is supported in all GSS packages, as well as packages like GeoChoicePerspectives. A spatial version of the one-person, one-vote approach is implemented in ActiveResponse GIS (Faber, Wallace and Miller 1996). In the spatial version, e.g. in the context of forest plan support, each participant can outline an area(s) to be protected (or not protected depending on the proposition under consideration). Software will perform the polygon overlay to compute the greatest common areal unit of vote (protection), i.e. the intersection of the most overlapping polygons.

Another approach to voting is a ranked vote, whereby several option lists, each list ranked within itself, are considered for synthesis. GeoChoicePerspectives uses a modified Borda technique to combine ranked alternatives (Hwang and Lin 1987, Black 1958). The Borda technique assigns ranks to decision alternatives based on the rationale that the higher the position of an alternative plan on the voter's list, the higher the rank assigned. The voting position of an alternative is determined by adding the ranks for each alternative from every voter using the Borda vote aggregation function. The winner is an alternative that receives the highest score calculated, such that all alternatives are assigned a score starting with 0 for the least favorable solution, 1 for the second worst, 2 for the third worst, and so on. All scores are weighted by the number of voters, resulting in the Borda score for each alternative. This type of vote aggregation prevents a contentious alternative, that ranks very high with some group members and very low with others, from winning, and promotes a consensus alternative. As an example, a consensus rank table and its corresponding consensus rank map based on Borda vote aggregation function are depicted in Figure 3.3 and Plate 3, respectively.

In Figure 3.3 low variance scores indicate situations where the three decision-makers prioritized an option in about the same position in the list in each of their individual lists ("0" indicates exactly the same position for all three). High variance scores are indicative of one decision maker ranking an option higher in a ranked-list relative to other decision makers who might have ranked the same site lower, or in the middle of the list.

In Plate 3 higher priority option sites (ranked as 3,5,7...) are depicted with larger circles, and lower priority option sites with smaller circles. The variance (from Figure 3.3) among the combined ranks is depicted using a color (shading) scheme. Green (showing as medium gray in a black and white image) represents more consensus, i.e. in the lower one-third of

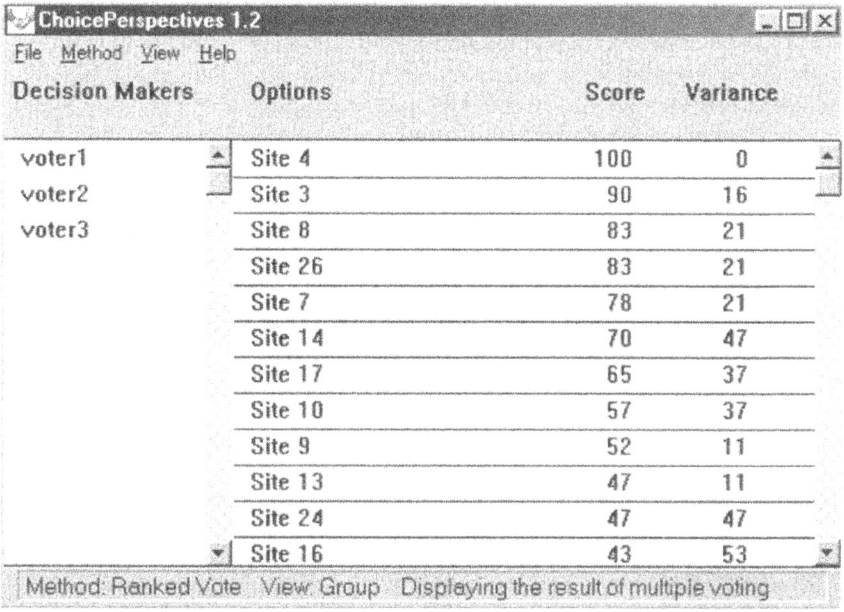

ChoicePerspectives 1.2

File  Method  View  Help

| Decision Makers | Options | Score | Variance |
|---|---|---|---|
| voter1 | Site 4 | 100 | 0 |
| voter2 | Site 3 | 90 | 16 |
| voter3 | Site 8 | 83 | 21 |
| | Site 26 | 83 | 21 |
| | Site 7 | 78 | 21 |
| | Site 14 | 70 | 47 |
| | Site 17 | 65 | 37 |
| | Site 10 | 57 | 37 |
| | Site 9 | 52 | 11 |
| | Site 13 | 47 | 11 |
| | Site 24 | 47 | 47 |
| | Site 16 | 43 | 53 |

Method: Ranked Vote   View: Group   Displaying the result of multiple voting

*Figure 3.3* Consensus scores and variances for habitat site options

variance scores, and red (showing as a darker gray if in black and white) represents less consensus, i.e. in the higher one-third of variance scores.

It is frequently argued that the Borda method and other point assignment methods are sensitive to manipulation of results by changing voters' agenda. Introduction of an "irrelevant" decision option to the system can effectively reverse the point-total order of existing options although no changes were made in voters' rankings. In fact, the original Borda method itself may violate the majority rule in that it can defeat an option that is the first choice of the majority. Recently proposed Borda Count derivative systems, including Borda Quota System and Borda Proportional Representation (Dummett 1997) were developed to counteract these problems. Gehrlein *et al.* (1982) argued that Borda is unlikely to change the winner when a losing alternative is being removed from the alternatives being voted on. Although for a long time the Borda score was overshadowed by the Condorcet social choice function, it seems once more to be gaining recognition, which can be illustrated by its frequent use in private companies and sport organizations (e.g. the Olympics, National Football League, or Tour de France) (Brams and Fishburn 1991).

### *Level 2: Decision analysis support*

*(d) Option modelling*  Options in spatial decision making are usually associated with locations. Finding suitable locations, i.e. locations that conjunctively satisfy locational criteria, has been a mainstay of modeling in GIS. Locational decision options are typically implemented in GIS by means of exclusionary screening procedures. These procedures include:

1  selecting locational criteria, e.g. median household income, density of road network, availability of vacant commercially zoned parcels, etc;
2  generating individual suitability maps for each locational criterion; and
3  combining suitability maps through either a Boolean overlay, wherein locations meeting each individual criterion are combined through spatial/logical operators (intersection overlay corresponding to logical AND, union overlay corresponding to logical OR), or weighted linear combination, where locational criteria values are standardized to a common numeric scale, and then combined by weighted averaging.

Since some locational criteria, e.g. landscape amenities, may be qualitative in nature and difficult to represent as crisp numeric requirements, *fuzzy set*-based screening procedures were recently suggested as an alternative approach to generating suitable location alternatives (Jiang and Eastman 2000).

Process models capture the dynamic character of phenomena in attempts to model the change in that phenomena over time, e.g. water flow changes in a river affected by pollution. Such models have been linked with GIS, and this linkage has been traditionally thought to be the essence of spatial decision support. Reitsma *et al.* (1996) provides insight into the structure and support of decision making using decision support system for water resource management. Reitsma *et al.* (1996) report on small-group use of a simulation model in a decision support system for water resource management negotiation focusing on water releases into the Colorado River from dams, but describe mixed results with their three-person groups due to the technical complexity of the model. Regarding group size, it is useful to point out that Vogel (1993) reviewed several experimental studies of GSS use and concluded that groups of five or more have a better chance of showing improvements in group process, whereas groups of three or four might show only negligible benefits from the introduction of GSS technology. Furthermore, the more complex the modeling activity, the more challenge there is to get participants to understand the dynamics of the process being modeled. The results of the study conducted by Reitsma *et al.* (1996), and another study with simpler spatial choice models (Stasik 1999), indicate the need for new metaphors helping the non-specialist participants, e.g. representatives of citizen groups, non-expert stakeholders, understand modeled processes and model results as they determine decision option trade-offs.

One example of such a metaphor is *decision maps* which depict trade-offs among key decision objectives embedded in a process model (Jankowski *et al.* 1999, Lotov *et al.* 2000). The purpose of decision maps is to help understand the relationships among decision objectives, which otherwise might be difficult to glean from a model. Decision maps greatly simplify the complexity of a model by replacing its often complex mathematical form with input–output dependencies. A matrix of input–output dependencies, called an *influence matrix*, is constructed prior to displaying a decision map. This step is accomplished by parametrizing a model, i.e. running a model many times against different input values and model parameters. An example of a decision map for a water pollutant transport model MIKE11 is given in Plate 4. MIKE11 is a two-dimensional hydrologic model used to simulate the concentrations of pollutants in a stream, using as inputs pollutant discharge data along with hydrologic and water quality characteristics (Danish Hydraulic Institute 2000). In Plate 4 the values of oil pollutant concentrations (Z5), computed with MIKE11, are given on the vertical axis, the water treatment costs (F) are given on the horizontal axis, and the value intervals of nitrates concentrations (Z4) are given on the color scale located below the horizontal axis. The decision map contains three slices: blue, corresponding to nitrates concentration equal to 2.5 units, dark blue equal to 2 units, and magenta equal to 1.5 units. Other slices (green, yellow, and red) represent nitrates concentrations higher than 2.5 units overlap and are barely visible in the upper left part of the map — along the top vertical axis.

The decision map in Plate 4 allows us to analyze trade-offs among nitrates, oil pollutants, and the cost of treatment by examining the frontiers (edges) of slices in the lower left side of the map. Moving from a zero point, along the cost axis (horizontal axis), to the right by a tiny cost increment one can change a slice from red (highest nitrates concentration of 5.5 units) to blue (equal to 2.5 units). This corresponds to reducing the nitrates in a river by more than one-half at a relatively small cost of about $20 million (see the horizontal axis in Plate 4 representing F ("cost of water treatment"). At the same time, reducing the concentration of oil pollutants from the maximum level of 3.401 to about 1.5 (location of the crosshair on the map) costs about $700 million. By analyzing further the frontier along the blue and dark blue slices, the decision makers/group participants can quickly learn about the major trade-offs of this decision situation. It is relatively less expensive to cut the oil pollutant concentration by one-half (from 3.4 to 1.7). Moreover, this reduction in oil pollutants can be done while simultaneously reducing the nitrates concentration from 5.5 to 2.5 (moving from red to blue slice). However, any improvement beyond 1.7, both in terms of nitrates and oil pollutants is going to be very costly.

Metaphors such as decision maps are needed if complex models of dynamic environmental processes are going to be embraced by non-expert decision participants. In this concrete example the decision map acts as the model summary. It encapsulates the results of many simulation runs in a form of a two-dimensional graph depicting the essential trade-offs among two indicators of water quality and the cost of water treatment. These trade-offs may provide the basis for selecting a particular water treatment alternative, hence making a decision.

*(e) Choice models*  Choice models provide assistance in comparing numerous options against each other in terms of criteria in order to select the best options. Multiple criteria decision making (MCDM) models are among the most popular of these, having been researched for at least 30 years, and implemented as software tools for at least the past 15 years (Ozernoy 1991). There are many kinds of multiple criteria decision models, each having advantages and disadvantages. Ozernoy (1991) presents a detailed discussion of using MCDM models to select the best MCDM methods.

Integration of multiple criteria decision models into GSS has been an active area of research for over 15 years (Jelassi, Jarke and Stohr, 1985), but few have found their way into commercial systems. The Meeting-Works software (Lewis 1994) is one that does have a MCDM model. However, because there are so many types of decision analyses techniques (hence models), an exhaustive study of which ones to include has probably prohibited most companies from investing in this development path. From the opposite perspective, Expert Choice Inc. is a company that has been incorporating GSS capabilities into its MCDM product. Team Expert

Choice implements a specific MCDM technique called analytical hierarchy process (AHP) in a group context, but does not provide a distributed group communication capability (Expert Choice Inc. 1994, 1995).

Integration of multiple criteria decision models with GIS has been an active area of research throughout the 1990s (Janssen and Rietveld 1990, Carver 1991, Jankowski 1995), and is now becoming a development activity. The IDRISI GIS software has MCDM models integrated in the package; however, the human-computer interface is still designed for individual use (Eastman *et al.* 1995). Jankowski (1995) has undertaken a thorough study of the advantages and disadvantages of various "aggregation techniques", creating a framework to understand their differences. From this basis of understanding, three aggregation techniques were implemented as part of GeoChoicePerspectives. The three techniques are *weighted summation, rank order*, and *ideal point*. Weighted summation is the most popular technique, as it is widely used because of its simplicity. Rank order is similar, but does not make the same assumptions about "interval level data" in the aggregation process, hence is somewhat more defensible. Ideal point is among the more logically appropriate because of its assumptions about preferred options. Each technique is briefly described below and an example using the same criteria with the same weights on the criteria is provided, so that the reader can form an understanding of how the nature of the computations change the nature of the evaluation scores, and hence the rank order of options.

The *weighted summation* technique is based on a linear combination of criterion data values and weights (Voogd 1983). Weighted summation is the most popular of the aggregation techniques because of its mathematical simplicity. Each criterion data value is normalized across the range of data values for that criterion in the database. The weights are also normalized to span a range of 1–100. A weighted criterion score is computed by multiplying each normalized criterion data value by the corresponding criterion weight. Criterion weight represents a priority/ importance that can be assigned to each criterion. An evaluation score for an option is calculated by summing the weighted criterion scores. The sum of the products calculated for each option represents the evaluation score for that option, all of the scores being normalized since the criterion data values and weights are all normalized (see Figure 3.4).

The *rank order* technique uses weighted summation with a linear combination of weights and scores, but with rank ordering as the basis of computing the evaluation scores. Instead of using the interval scale properties of the criterion scores, each alternative's criterion score is based on the alternative's position in the ordered list of all of the scores for a criterion across all alternatives. Since the scores are normalized for each criterion, an interval level of measurement exists. The ranks (normalized scores) are then summed for each alternative, resulting in a rank ordered score between 1 and 100 (see Figure 3.5).

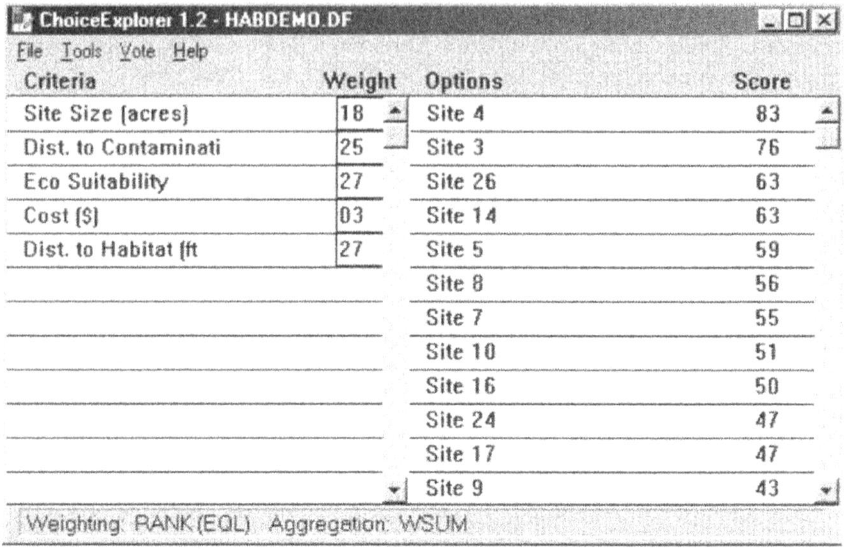

| ChoiceExplorer 1.2 - HABDEMO.DF | | | _ □ × |
|---|---|---|---|
| File  Tools  Vote  Help | | | |
| **Criteria** | **Weight** | **Options** | **Score** |
| Site Size (acres) | 18 | Site 4 | 83 |
| Dist. to Contaminati | 25 | Site 3 | 76 |
| Eco Suitability | 27 | Site 26 | 63 |
| Cost ($) | 03 | Site 14 | 63 |
| Dist. to Habitat (ft | 27 | Site 5 | 59 |
| | | Site 8 | 56 |
| | | Site 7 | 55 |
| | | Site 10 | 51 |
| | | Site 16 | 50 |
| | | Site 24 | 47 |
| | | Site 17 | 47 |
| | | Site 9 | 43 |
| Weighting: RANK (EQL)  Aggregation: WSUM | | | |

*Figure 3.4* Weighted summation technique applied to habitat site evaluation

| ChoiceExplorer 1.2 - HABDEMO.DF | | | _ □ × |
|---|---|---|---|
| File  Tools  Vote  Help | | | |
| **Criteria** | **Weight** | **Options** | **Score** |
| Site Size (acres) | 18 | Site 4 | 89 |
| Dist. to Contaminati | 25 | Site 3 | 85 |
| Eco Suitability | 27 | Site 26 | 79 |
| Cost ($) | 03 | Site 14 | 78 |
| Dist. to Habitat (ft | 27 | Site 5 | 64 |
| | | Site 7 | 61 |
| | | Site 13 | 58 |
| | | Site 17 | 58 |
| | | Site 16 | 57 |
| | | Site 8 | 49 |
| | | Site 9 | 48 |
| | | Site 10 | 46 |
| Weighting: RANK (EQL)  Aggregation: RO | | | |

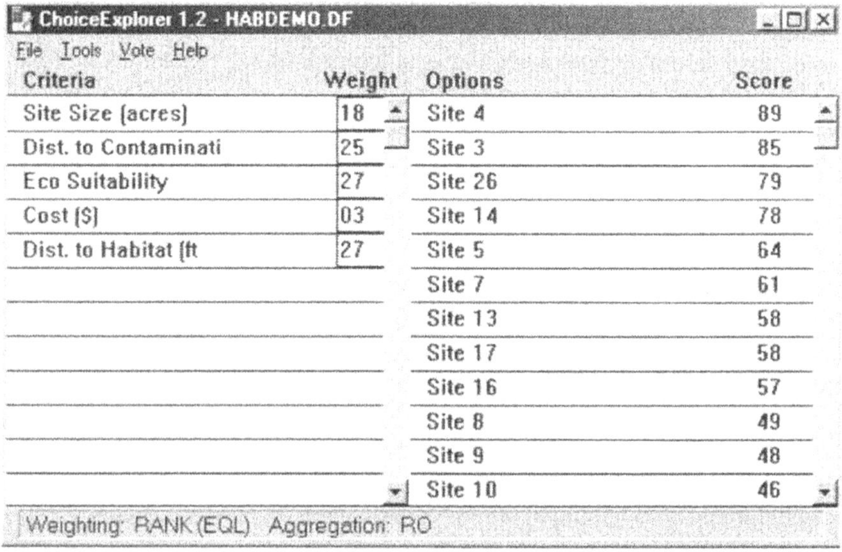

*Figure 3.5* Rank order technique applied to habitat site evaluation

The difference between the option scores of Figures 3.4 and 3.5 is based on the different way that weighted criterion scores are scaled, using interval scale in the case of weighted summation (Figure 3.4) and an ordinal scale in the case of rank order (Figure 3.5). The difference is due to the criterion "Dist (distance) to habitat" as measured in feet. Distances become ranks of distance in the rank order aggregation method, rather than interval measurements as in the weighted summation. Consequently, the rank approach understates the differences in distance. Alternatively, differences in an ordinal measurement (1–5) as for "public access" are overstated when compared to other interval measured data. What this means is that one should use the rank order aggregation technique when most of the criteria are measured as "ordinal ranks", and one should use the weighted summation approach when the criterion values are measurable on an interval/ratio scale.

As a third approach, the *ideal point* is more similar to weighted summation than it is to rank order. Ideal point makes an assessment for each criterion in the database in regard to which data value is the most preferred (using an array of values representing the observed or predicted performance of all considered options), and which data value in the database is the least preferred. In a sense, the technique constructs an "ideal option" by scanning all criterion data values for all options. In the same sense, the technique constructs a "least preferred option". It uses these options, i.e. the criteria associated with the options, as anchors to compute weighted data values for each criterion, and then sums those weighted data values to compute the option score in the same manner as the weighted summation method (see Figure 3.6).

*(f) Structured-group process techniques*  Agendas help structure group interaction for meetings; but agenda topics do not help structure conversation about those topics. The structuring is called "group process". GSS have been developed to help reduce what is called group process loss, i.e. loss of productivity due to "wandering" social interaction. For several years techniques of various kinds have been available to help facilitate group interaction. Some of the better known ones that can be used for structuring computer-assisted meetings, presented in the order of their structuring rigor, are: brainstorming, Delphi, modified Delphi, and technology of participation. Electronic brainstorming proceeds with a facilitator requesting participants to contribute simultaneously ideas to a topic. Those ideas are commonly displayed as they are submitted, and the author is made known (Hwang and Lin 1987). The Delphi technique proceeds with a facilitator requesting information of participants in secret, i.e. noone tells others who authored what information. The idea elicitation process is iterative, in that several rounds of request and contribution are commonly performed, with the results of each round being synthesized and then resubmitted to the participants. This technique has been popular in conflict laden meetings, where the "power" of participants is to be controlled so that everyone can contribute on an equal level (Hwang and Lin 1987). The modified Delphi is similar to Delphi except that authorship of ideas is known. This technique has been useful in group meetings where

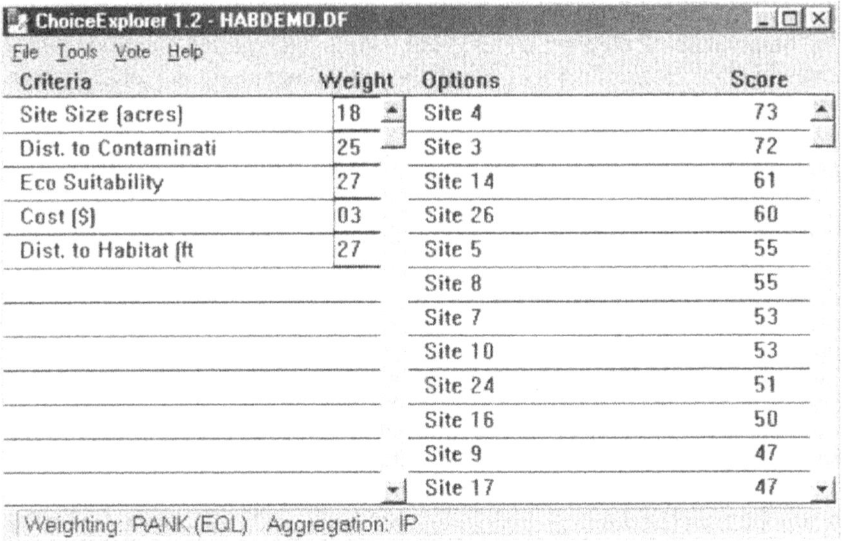

| Criteria | Weight | Options | Score |
|----------|--------|---------|-------|
| Site Size (acres) | 18 | Site 4 | 73 |
| Dist. to Contaminati | 25 | Site 3 | 72 |
| Eco Suitability | 27 | Site 14 | 61 |
| Cost ($) | 03 | Site 26 | 60 |
| Dist. to Habitat (ft | 27 | Site 5 | 55 |
| | | Site 8 | 55 |
| | | Site 7 | 53 |
| | | Site 10 | 53 |
| | | Site 24 | 51 |
| | | Site 16 | 50 |
| | | Site 9 | 47 |
| | | Site 17 | 47 |

ChoiceExplorer 1.2 - HABDEMO.DF
File  Tools  Vote  Help
Weighting: RANK (EQL)  Aggregation: IP

*Figure 3.6* Ideal point technique applied to habitat site evaluation

expert opinion is used to resolve uncertainty about environmental problems (Webler *et al.* 1991), and where large groups of people have taken part in participatory decision making (Renn *et al.* 1993). The technology of participation (ToP) makes use of a four-step process for strategic planning that includes idea (issue) elicitation, clustering of ideas into similar topics and identifying constraints, establishing priority (direction) in line with constraints, and writing a plan to carry out the prioritized issues (Spencer 1989). The ToP process has been used in strategic planning efforts around the world, but has not found its way into GSS software as yet, to our knowledge.

Group process techniques, together with the techniques for collaborative communication support, can be used to organize meetings for structured creativity. Such techniques are often useful at the beginning of group processes, where information creativity and planning tasks are more common, than at the end of group processes since conversation at the end is more directed. Many of the GSS described by Bostrom, Watson and Kinney (1992) implement such capabilities, making use of a facilitator to encourage group process direction. The Active Response GIS (Faber, Wallace and Cuthbertson 1995) in its linkage with Group Systems for Windows (GroupSystems 2000), can make use of electronic brainstorming to collect ideas for plan making.

### Level 3: Group reasoning support

*(g) Judgement refinement/amplification techniques* Judgement refinement and amplification techniques are specialized techniques for detailing the character of choices made in relation to the overall pattern of choices. One of the most significant developments has been the addition of sensitivity analysis to MCDM packages. Expert Choice (Expert Choice Inc. 1994) is one such package that has had this, and now implemented sensitivity analysis capabilities for group support in a software package called Team Expert Choice (Expert Choice Inc. 1995). GeoChoicePerspectives links sensitivity analysis techniques with map displays to visualize the spatial manifestations of alternative rankings of options. During sensitivity analysis a criterion weight can be changed relative to other weights, e.g. in GeoChoicePerspectives the importance of criteria in relation to each other influences the priorities of habitat sites (see Figure 3.7(a) and (b)). The ideal point aggregation method (in Figure 3.6) was used as a basis for the score computation. Figure 3.7(a) shows equal weights on the left with the corresponding rank order indicated by bar length (longer bar is higher rank). Figure 3.7(b) shows two weight changes and the corresponding changes in the rank order (on the right side of the window).

Together, representation aids such as maps and sensitivity analysis can be used in judgement refinement, as available in GeoChoicePerspectives (see Figure 3.8(a) and (b)). Figures 3.8(a) and 3.8(b) show graduated circle maps depicting habitat site rankings according to the criteria weights in Figure 3.7(a) and 3.7(b), respectively. Larger circles indicate higher ranks and smaller circles indicate lower ranks. Clearly, different types of maps could be used, but each different map type must have a fundamental advantage to it. In the case of graduated circles, the visual variable "size" is manipulated. Size is one of the easiest visual cue changes to detect for the human eye — even more so than shaded gray tones that are common on choropleth maps.

Software packages implement Bayesian statistical techniques for decision analysis and judgement refinement (Bayesian Systems Inc. 2000). Bayesian inferencing is based on probability theory. Bayesian statistics focus on incremental knowledge improvement to reduce uncertainty when considering trade-offs among criteria for prioritizing options. Since Bayesian inferencing is a relatively new development, it has not yet found its way into GSS or GIS packages, with the exception of IDRISI GIS software and an extension developed for Arcview GIS software.

*(h) Analytical reasoning techniques* Analysis and reasoning techniques are still research topics for the most part. Expert systems and mathematical programming packages are examples of such techniques that have been integrated with GSS on a limited basis (Jarke 1986). Research prototypes attempt to track the decision process using syllogistic reasoning. Bayesian inferencing engines can help with tracking decisions

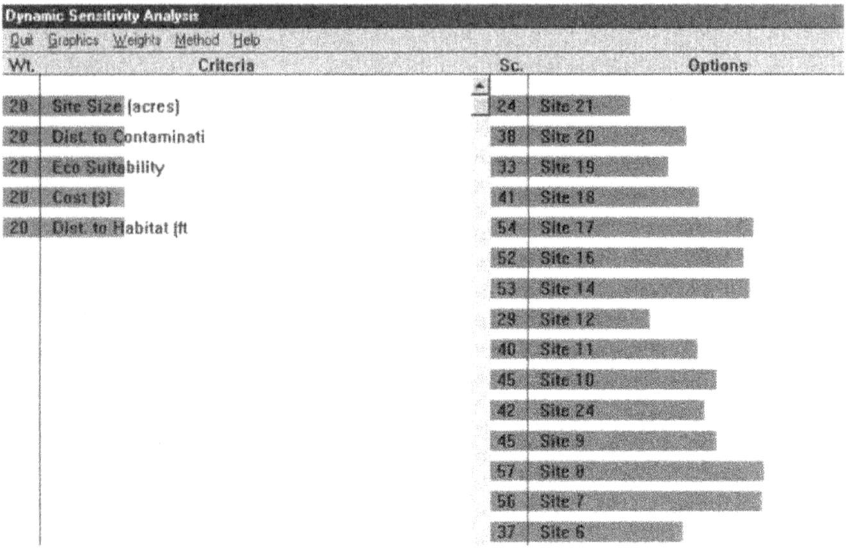

*Figure 3.7a* Sensitivity analysis before weighting change

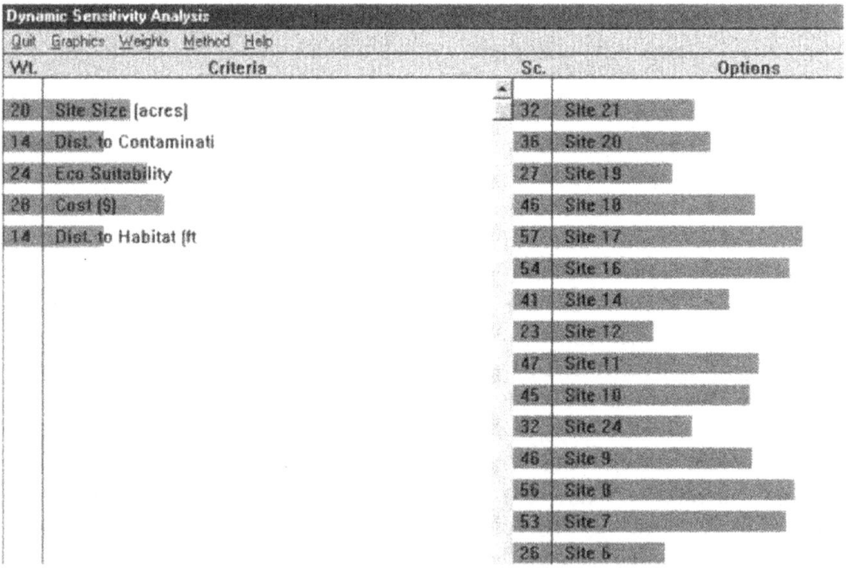

*Figure 3.7b* Sensitivity analysis after weighting change

for operational and structured kinds of decision problems, i.e. where routine inputs are used to reason, as in minor infrastructure repair (Bayesian Systems Inc. 2000). A promising approach to discovery of individual and group decision knowledge is *soft computing*. The basic thesis of soft computing is to imitate the human mind in exploiting the tol-

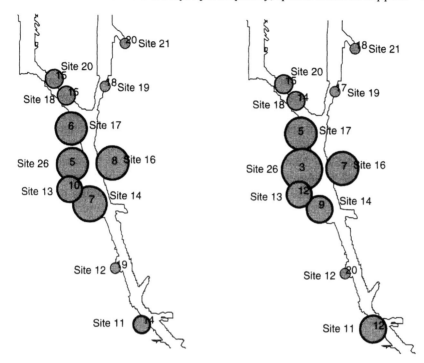

*Figure 3.8a* Rank map of habitat sites associated with criterion weights in Figure 3.7(a)

*Figure 3.8b* Rank map of habitat sites associated with criterion weights in Figure 3.7(b)

erance for imprecision and uncertainty when dealing with very complex and often imprecisely formulated tasks. The aim of soft computing is to achieve tractability and robustness in solving complex and imprecisely formulated problems. Methodological components of soft computing include, among others, fuzzy set theory, fuzzy logic, neural networks, and evolutionary algorithms. One example of a soft computing methodology applicable to repetitive decision situations is *rough sets theory* (Pawlak and Slowinski 1994). The rough sets methodology produces a set of decision rules involving a reduced number of most important decision criteria, hence diminishing the uncertainty inherently associated with criterion data values. The set of decision rules explains the decision situation and may be used to support new decisions.

An example of a research prototype integrating GIS with data mining — another component of soft computing — has already been mentioned DECADE (**D**ynamic, **E**xploratory **Ca**rtography for **De**cision Support) (Jankowski, Andrienko and Andrienko 2001). DECADE uses GIS database, knowledge and preferences of group participants and C4.5 classification tree derivation algorithm (Quinlan 1993) to discriminate among some given classes of objects and produce their collective descriptions on the basis of values of attributes associated with class members. The aim of this approach is to help group participants reduce the decision complexity of a problem by identifying key decision criteria.

Expert systems and mathematical programming packages have been integrated successfully with GIS for individual users, but group-based GIS integration is only beginning to be proposed as related to intelligent transportation system technology (Kanafani, Khattak and Dahlgren 1994). Such implementations rely on substantial information technology infrastructure, set up to collect substantial amounts of traffic data to make operational decisions. More investigation must be done to determine how useful analytical reasoning might be for strategic or managerial decision environments, which tend to shy away from complex information tools.

## 3.4  Personnel requirements

Besides the participants that are involved in a collaborative effort, there are two important roles in a collaborative setting that must be supported. The role of a facilitator is to provide decisional guidance in terms of steps in the decision making process (Schwarz 1994). A facilitator guides participants through a process, and keeps participants from threatening each other when the conversation gets heated. The role of a chauffeur is to provide advice to the participants on the use of software, hardware, and/or data capabilities during the use of the system. The former tends to be a social-oriented role, whereas the latter tends to be a technical, computer-oriented role. In their review of over ten years of experimental studies with decision groups, Chun and Park (1998) found that group performance, represented by decision time, depends on the familiarity of participants with decision support tools. A skilful group facilitator/chauffeur, however, can compensate for the lack of participant experience with the tools. Additionally, participant attitudes, measured by satisfaction with group decision support tools, depended strongly on the presence or absence of a facilitator. The presence of a facilitator enhanced participants' satisfaction with computer-supported decision process.

Sometimes two people are needed to support the two roles, but at other times only one person might be necessary. The difference usually is in the size of the group and the technical support needed. Larger groups (15 or more people) and/or more technical, software capabilities commonly encourage two people, each taking on one of the roles. Thus, for smaller groups and less technical support, only one person might be necessary.

## 3.5  Architectures for implementing collaborative decision support systems

Hardware and software architectures for collaborative decision support systems differ depending on the meeting arrangements (described in section 3.2.1). Based on the hardware considerations (described in section 3.2.2) two different architectural configurations can be implemented: low band-based (appropriate for same time–same place; different time–same place collaboration) and broad band-based (same time–different place; different time–different place collaboration).

Furthermore, each of the two hardware architectures can be developed in either pier-to-pier configuration or client-server configuration. A pier-to-pier configuration consists of two or more workstations communicating with each other over a network communications channel, but there is no database server servicing the network. A client-server configuration consists of one or more client workstations and one or more database servers connected over a data communications channel, with the client workstations drawing data from the server. The choice of configuration depends on how the data are managed. For minimal volume data management requirements, such as the collection of decision maker votes, pier-to-pier configuration will suffice. Small data management tasks can then be performed by a human facilitator/chauffeur. The client-server configuration becomes more important when the data management plays a larger role in supporting the decision making process. The server can manage data processing, access control, and communication tasks with- or without the participation of the facilitator/chauffeur.

### 3.5.1 Hardware architecture for same-place, same/different-time collaboration support

A low band-based architecture is appropriate for collaborative work occurring in the same place, where the notion of "same place" is understood as one meeting room. A conventional meeting is the most common form of collaborative decision making supported by this architecture. A low band-based architecture is also conducive for the computerized support of collaborative work during different time (distributed meeting). An example of the latter arrangement may be a land use planning process during which multiple participants have a certain amount of time, usually exceeding one day, to analyze data, propose planning options, evaluate option outcomes, vote on different options, and negotiate the choice of the most satisfying option.

One version of the low band-based architecture is a pier-to-pier configuration. A more complex version of this architecture involves the use of a local network or an intranet server (usually a higher end PC) equipped with server software (Figure 3.9). In this architecture, the local network server plays the role of data repository and carries out data management tasks, such as checking the authorization for data access and facilitating the communication among the participants.

Pier-to-pier connectivity among PCs in a local network works well for meetings taking place in the same room. More distributed collaborative

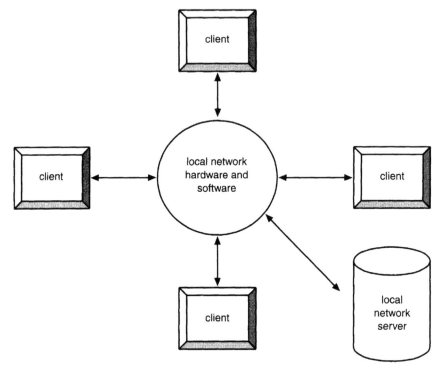

*Figure 3.9* Local network server-based architecture for computerized support of same-place collaborative decision making environment

work, e.g. taking place in different rooms of the same building, is better served by computers connected through Ethernet infrastructure. This requires that each PC be equipped with the Ethernet network card.

### 3.5.2 Software architecture for same-place, same/different-time collaboration support

The initial consideration for a software architecture is whether software is going to be implemented as one multifunctional tool or whether it is going to be comprised of various integrated software tools. The advantage of the former solution is seamless data transfer and unified user interface. The disadvantage is the development cost. Conversely, the advantage of integrated solution is lower development cost and its disadvantage is dissimilar user interfaces.

Another consideration for system design involves a specific hardware architecture solution. In the server-based solution the software can be developed using *client-server* software architecture. The main rationale for such architecture is the potential efficiency of information processing that can be realized from dividing processing functions between client and server, and data management and security that can be automatically controlled by the server. In the local network solution without a server, there is still the possibility of allocating certain functions of the software to be performed on individual participant machines and some functions to be carried out only on the facilitator/chauffeur machine, which does not have to be the server.

Architecture for a system to support same-place, same/different-time meeting arrangements can be described using an example from our efforts related to a prototype software implementation called Spatial Group Choice (Jankowski *et al.* 1997). Spatial Group Choice was developed for repeated group decision making experiments, a joint project between the Geography Department at the University of Idaho and the Geography Department at the University of Washington. Spatial Group Choice links two different software tools, hence it is an example of the integrated software architecture.

The hardware configuration of Spatial Group Choice included a network of PCs, large public screen, and an LCD panel with a high intensity overhead projector connected to a facilitator/chauffeur computer (the inner "U" in Figure 3.2). The software components include operating system, network software, multiple criteria evaluation software for decision modeling and voting, and interactive map visualization software. The software architecture of Spatial Group Choice is presented in Figure 3.10.

Spatial Group Choice is comprised of two modules: a multiple criteria evaluation software called Group Choice, with a submodule for multiple

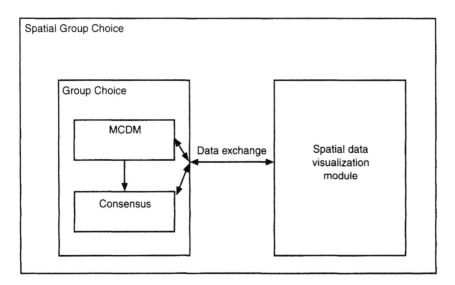

*Figure 3.10* Software architecture for Spatial Group Choice based on software module integration through dynamic data exchange protocol

criteria decision making (called MCDM) and a voting submodule (called Consensus), and a module for spatial visualization, customized from ArcView software produced by Environmental Systems Research Institute (ESRI) Inc. Consequently, the integrated software of Group Choice together with ArcView 2.1 is called Spatial Group Choice. The software integration is based on a loose-coupling strategy (Jankowski 1995).

Spatial Group Choice can be used in two modes, individual and group. In the individual mode the MCDM program has tools and methods to assist an individual group member in the evaluation of decision alternatives and the submission of his/her opinion to the whole group. The submission is done through a vote and can be open or anonymous. In the group mode the Consensus program has tools to calculate and display voting results, including a measure of group agreement/disagreement. Both programs (MCDM and Consensus) can be used in the group mode using a large-screen computer display.

The spatial data visualization module can be used in both individual and group modes, and serves as the database repository and data visualization tool. The decision problem database is organized in a project directory comprised of coverages, and orthophoto image files. The problem-related data is organized in an ArcView 2.1 project in themes where every theme is represented by a customized thematic map. For the support of problem solving using multiple criteria evaluation, the visualization module provides graduated symbol-based maps and histoframe thematic maps. These maps enable the comparison of different decision criterion values and their spatial distribution. The thematic maps for a problem-specific ArcView 2.1 project can present either decision criteria or background information. These and other capabilities were implemented in one of the first commercial software products for collaborative spatial decision making support, called GeoChoicePerspectives available from Geo Choice Inc. (Geo Choice Inc. 2000).

### 3.5.3 Hardware architecture for different-place, same/different-time collaboration support

A popular form of collaborative decision making within a different-place arrangement is a teleconference, where the participants are distributed in different locations and the medium of communication is a television connecting the participants via a closed-circuit or a satellite link. Another form, albeit limited to audio communication, is a multiparty telephone conference, where participants become simultaneous callers to a conference.

These forms of supporting group decision making are suitable for *different-place same-time* collaboration. They are also limited in data management and analytical support functions. For these reasons, computerized support has been developing as a new technology that (given sufficient bandwidth) can provide video and audio capabilities plus data management and analytical support, otherwise unavailable in conventional teleconferencing. The internet has been a touted communication infrastructure on which the computerized support for collaborative work can be developed. Although at the current bandwidth the internet infrastructure is mostly incapable of supporting real-time video and audio, its communication capabilities are sufficient to support data management and analytical functions necessary for collaborative spatial decision making.

Although one could envision the internet-based architecture without a server, a practical solution for most space and time distributed decision support systems will involve a client- server architecture (Figure 3.11).

The server can be any high-end PC or Unix workstation with built-in hardware support for network connectivity. It is important that the server be connected to the internet infrastructure via a high-speed connection operating at a speed of at least 1 megabit per second. Clients can be connected to the internet either via a dedicated line, a high-speed modem connection (e.g. DSL or cable modems) or a wireless connection.

### 3.5.4 Software architecture for different-place, same/different-time collaboration support

There are three approaches that one can consider when designing a spatial decision support system with internet tools (web server and web browser). First, from the client software perspective (web browser), an application may work as an *applet* (mini-application) written in Java programming language (Stasik 1999, Jankowski and Stasik 1997). Java applets are pre-compiled, working "inside" a web document, and provide enhancements to a static web page content. This approach can be considered for the internet-based decision support capabilities that require limited (small-size) GIS applications. Many GIS applications, however, are large and therefore take time to download. Since Java applets have to be re-downloaded every time users revisit/refresh a web page, the time needed to load applets becomes an issue. This limitation currently constrains the use of Java applets to a few simple spatial analysis functions useful for

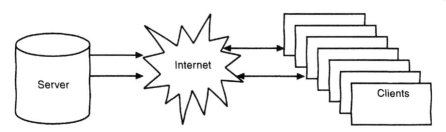

*Figure 3.11* Hardware architecture for distributed space and time collaborative decision support system

decision support (e.g. distance-based query, point-in-polygon search, buffer and overlay).

An alternative solution to the above problem involves creating GIS software that functions like a web browser component. Examples include numerous *plug-ins* to the Netscape Navigator web browser. A plug-in is an independent program that works in the frame of a browser, placing itself between a browser and data to be accessed. A fundamental difference between a plug-in and a Java applet lies in how they are accessed by the user. A plug-in must be initially downloaded from a server to the user machine's hard disk, after which it is accessed from the hard disk. This enables a plug-in to access other files on the hard disk. A Java applet, in contrast to a plug-in, is loaded every time from a server and it exists only in the context of a specific WWW page that uses it. The advantage of the plug-in solution over the Java applet is that in the case of complex software, such as GIS, the user does not have to send data manually to an external program. The web browser does it automatically.

Although a design of collaborative spatial decision support system based on web browser and plug-in architecture seems to be a promising solution, it is also not free of limitations. The most severe is the dependence on web browser software. A plug-in may need to be implemented specifically for a given web browser. This means that there would have to be a separate plug-in for Netscape Navigator and for MS Internet Explorer. If a new version of browser software becomes available a plug-in must be updated. In addition, sending a large volume of data from a web browser poses another limitation. Large data transmission tends to slow performance, reducing the sense of interaction in an application.

The third approach involves the development of independent, specialized software that allows a client to communicate with the server using the internet infrastructure. An example of such architecture is prototype software for spatial understanding and decision support system (SUDSS) (Stasik 1999, Jankowski and Stasik 1997). SUDSS is based on the client-server architecture (see Figure 3.12). The server component, working under the MS Windows NT 4.0 operating system, handles the database management tasks. The client (user) component, working under the MS Windows 95 operating system, is a stand-alone application, which communicates with the database on the server. The data exchange between SUDSS client software and the SUDSS server is solely based on the TCP/IP protocol. TCP/IP-based communication provides reliable client-server connectivity using a packet-switched data transmission. A potential weakness of the protocol comes from the fact that it may operate on top of lower level, less reliable protocols. This may present a problem in case of multimedia transmission (especially live). However, the TCP/IP service was sufficient for the SUDSS prototype and its data management and analysis capabilities.

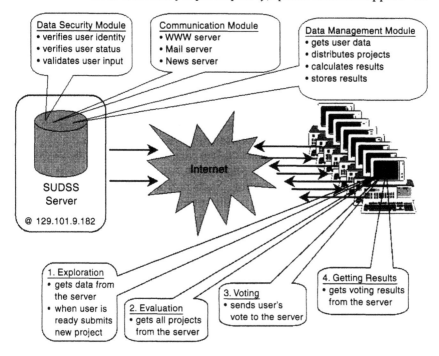

*Figure 3.12* Software architecture of SUDSS prototype

The advantage of the approach used in SUDSS prototype (a stand-alone, internet-aware software) is its independence from a web browser and better data management capabilities. In addition, since much of analytical functions and data management can be performed on the client, this reduces the network communication time, and hence speeds up software operations. The disadvantage of this solution is the development cost and often new "look" and "feel" of the user interface, which is different from what many users may be accustomed to in popular web browsers.

## 3.6 Examples of GIS-supported collaborative decision making software packages

Several GIS products exist that are just beginning to address group-based decision support. Among them are INDEX® from Criterion Inc., Smart-Places E from the Electrical Power Research Institute, and GeoChoice-Perspectives™ from Geo Choice Inc. Commercial products tend to lag behind advanced developments in distributed data communications covered in previous sections; however, the rich set of capabilities make these packages functional for supporting same-time and same-place meetings.

### 3.6.1 INDEX®

Criterion Inc. of Portland, Oregon offers consulting services to customize an application in ArcView GIS called INDEX®. INDEX® helps answer questions about sustainable development and community liveability. Such questions include the following. How can communities cope with growth and still maintain their quality of life? Are incremental development decisions making conditions more or less sustainable? How can services be most effectively focused in redevelopment areas?

Criterion makes a point of stating that INDEX® is not a "shrink-wrap" application, but a customizable application for developing quality of life indicators in a focus group setting. Such indicators can describe what a community feels is their current quality of life, and then create a plan that describes a (future) desired quality of life. The plans take the form of ArcView GIS maps. Criterion has offered the customization strategy since 1995; it appears to work well for them as a consulting firm. The customization comes in the form of data available and the region to which it is applied. The richness of problems and applying the software encourages the consulting strategy associated with this product.

INDEX® runs as an extension of ArcView GIS. Stakeholders can perform spatial accounting with local indicators to gauge not only what the positives and negatives are, but also where INDEX® can be applied to a single neighborhood or an entire region with a range of measurements that includes land-use, housing, employment, transportation, infrastructure, and the natural environment. Its multimedia functions can help engage citizens, translating abstract objectives into backyard tactics that can be visualized. INDEX® does not produce the perfect solution to community choices, but it is a powerful planning and decision-support tool that can improve the inclusiveness and accountability of community development processes.

### 3.6.2 *SmartPlaces E*

SmartPlaces E is a geographic decision support system that assists communities in assessing the implications and opportunities of alternative development plans (Electrical Power Research Institute 1998). Smart-Places offers innovation to the time-honored process of land use planning and evaluation through interactive decision making. SmartPlaces enhances decision maker insight for target marketing, economic development, land use planning, transportation systems, facilities management, environmental remediation and protection, energy forecast, water allocation and resource control. SmartPlaces scenario driven analysis evaluates the implications and opportunities of alternative strategies. SmartPlaces is a framework, which can be quickly applied to small rural towns as well as large urban environments.

The SmartPlaces system supports community planning through interactive review, plan experimentation, and impact evaluation models. It was designed with adjustable menus and selectable formulas to support a wide range of user interest and expertise. SmartPlaces software was developed on top of ESRI's ArcView geographic information systems (GIS) software. Using the ArcView GIS base, the SmartPlaces "Avenue" language product provides software to design, illustrate and evaluate land development scenarios. Parallel user screens for land-base information and decision-base information enable discovery of decision relationships.

Menu levels encourage diverse user skills, making the application approachable for new users, accessible for experienced users, and completely modifiable for program developers. There are four expertise levels available for the program. "New User" is for someone using the program for the first time, or with minimal experience in using the program. "Experienced User" is for someone who has considerable experience in using the program. "ArcView User" is for an experienced user who also has experience in the use of ArcView. "Manager" is for a developer or other technical person who has the capability to make changes in the program code.

SmartPlaces is an open system. The system provides four types of interactive computer tools for land use design and evaluation:

1   explore a design;
2   create and modify scenarios;
3   evaluate environmental and economic impacts of a proposed design; and
4   save a scenario for reference.

In SmartPlaces the territory or boundaries of the strategy are defined, the characteristics of the target area are identified, and the choices for analysis are selected. The user defines the issues and their related indicators, the measures for evaluation, the relative weight or importance of each criterion, the locations to be included in the analysis and the presentation method for the outcomes. The outcomes of each strategic choice are reported in both text format and graphic charts. Each set of definitions may be saved as a scenario and may be compared with other scenarios.

There are more than 300 Avenue scripts in SmartPlaces, comprising about 25,000 lines of Avenue code. Some internal functions are encrypted to protect product integrity. The system is open at all the connection points to user interface, scenario design, model evaluation, graphic presentation, and user functions for custom program implementation.

Key work functions are divided between a dual user interface. One window is devoted to the Scenario Builder view and its menu. The companion window is devoted to the Radix and its menu. Scenario Builder and Radix functions are listed with their user choices in the users' manual offered in the SmartPlaces home page (Electrical Power Research Institute 1998). Load and execute for each set of functions can be alternately accessed from the opposite GUI.

Routine Radix evaluation functions are driven by wizards which offer choices for how and what gets calculated with respect to the pocket attributes of the scenario. Custom Radix evaluations may be performed by doing system calls to companion analysis software executables, creating packaged evaluations in Avenue or calling functions written in another object language like Visual Basic or others preferred for model program work.

### 3.6.3  *GeoChoicePerspectives*<sup>TM</sup>

GeoChoicePerspectives™ (GCP) software, developed and marketed by Geo Choice Inc. of Redmond, Washington, USA (http://www.geochoice.com) supports single user and/or group-based decision making in a geographic information system context. Decision participants use GCP to explore, evaluate and prioritize preferences on all aspects of a decision making process involving multiple criteria and options. Options are represented as ArcView® GIS points, lines or areas with attributes. Multiple perspectives on options evaluation can be combined to provide an overall perspective. Single users can use the GCP to collate multiple evaluations of an option ranking. Groups can use GCP to combine multiple perspectives on criteria and options in an iterative process for consensus building.

The GCP package is composed of three components: GeoVisual™, ChoiceExplorer™, and ChoicePerspectives™ (see Figure 3.13). The GeoVisual™ component is used by decision participants for exploring geographic data on maps, and presenting the results of site options rankings for single user and/or group contexts that are generated by ChoiceExplorer™ or ChoicePerspectives™, respectively. GeoVisual™ is implemented as an extension of the ArcView® GIS platform. The ChoiceExplorer™ component is used by decision participants to perform criteria selection and weighting, plus options evaluation and prioritization. ChoicePerspectives™ collates rankings from ChoiceExplorer™ that are subsequently displayed as consensus maps in GeoVisual™. GeoVisual™ and ChoiceExplorer™ are dynamically linked to support interactive computation and display.

The GeoChoicePerspectives™ package can be used in a variety of meeting arrangements:

- in face-to-face meetings — participants meet at the same place and same time;
- in storyboard meetings — participants meet at the same place, but at different times;

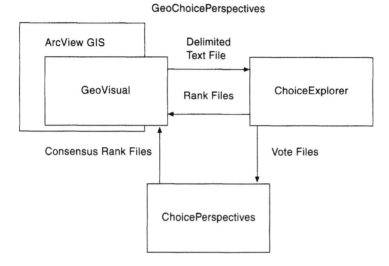

*Figure 3.13* GeoChoicePerspectives software architecture

- in conference call meetings — participants meet in different places at the same time;
- in distributed meetings — participants meet in different places at different (convenient) times.

Single copies of GeoChoicePerspectives™ can support face-to-face and storyboard meetings (i.e. same-place meetings). Multiple copies are needed to support conference call and distributed meetings (i.e. different-place meetings).

## 3.7 Conclusion

Methods and tools for participatory spatial decision making, just like methods and tools used in GIS, come from many disciplines, including cartography, computer science, operations research, psychology, cognitive science and management information science. In order to guide the review of methods and tools we used a normative model of decision making process called macro-micro decision strategy. According to this model the decision process can be represented by three phases: intelligence, design and choice (columns of a matrix), where each phase takes place during the course of four activities: gather, organize, select and review (represented by the rows of a matrix). Using the matrix format we assigned to each cell in the matrix a specific method/tool, which we identified as potentially useful for supporting a specific phase-activity. These methods/tools include information management and structured-group process techniques, representation aids, group collaboration support methods, process models, choice models, and judgement refinement techniques. How these methods and tools can be deployed to support participatory decision making depends on the type of participants, e.g. novices or experts, and meeting arrangements for collaborative work, e.g. face-to-face meeting, long-distance conference, different place/different time group work.

Some or all of the decision aiding methods and tools could potentially be useful for developing a participatory GIS. In order to provide guidelines for such a development we listed the methods and tools in order from the most basic to the most sophisticated, and treated them as building blocks of a system. At the most basic level called "basic information handling support" we distinguished information management, visual aids and simple group collaboration support, including generation and compilation of ideas. Information management tools allow the participants to access a problem-specific database(s) through map-based and attribute value-based queries. More advanced tools include group-based database design languages, such as Unified Modeling Language. Even though such languages might be used more aptly by decision analysts rather than decision makers themselves, it makes sense to have a group review of the basic data categories that will be used to frame a decision problem. Visual aids are another basic building component of participatory GIS. They include multimedia support, as indicated by photo and sound manipulation capabilities, charts, diagrams and tables linked to maps to enhance information presentations, and animation and virtual reality representations to enhance the cognitive experience of participants. The third basic component of participatory GIS, group collaboration support tools, include data and voice transmission, electronic voting, electronic white boards, computer conferencing, large-screen displays, computer-based idea generation, collection, display, discussion, and voting. Simultaneous and anonymous idea generation from any computer linked in a group is one of the major efficiencies brought about by group collaboration support tools. Sorting through the ideas and then placing them in categories assists in synthesizing ideas. The synthesis by the group leader/facilitator can then be redisplayed to all participants in the group. Participants can vote on lists of ideas and adopt the ones that garner the broadest support.

At the second level (in the order of increasing sophistication) called "decision analysis support" we included option generation techniques, such as GIS-based suitability modeling, process models generating alternative futures, i.e. simulating the outcomes of decision options on a number of performance indicators, choice models used to evaluate and order various decision options from the most to the least desirable, and structured group process techniques, including computer-supported brain storming and the technology of participation. At the third level called "group reasoning support" we included judgement refinement techniques such as the sensitivity analysis of choice model results and analytical reasoning techniques, including fuzzy logic, rough sets, and data mining.

The reader should note that the first level of capabilities is fundamental to facilitating interaction among participants, i.e. promoting the essence of "participatory activity". The state of technology development is developed more at level 1 than for levels 2 and 3, as evidenced by commercial product offerings. This is to be expected because these capabilities require the least customization for each decision situation — hence the market is larger at this level than the other levels. In the future we expect to see continued growth in product offerings by various software vendors.

The second level of capabilities builds on that first level, but not just by addition, but by integrating the two sets. Tight coupling of information management, display and communications capabilities at the first level with analysis, choice, and group-process structuring of the second level are needed to provide for seamless interaction among participants. The seamless interaction becomes more important when complex problems become ever more idiosyncratic to place-based and process-oriented discussions. Loosely coupled capabilities between the two levels is likely to create frustration among participants. Unfortunately, the tighter the integration, the more costly such capabilities are to provide — hence they are less likely to be provided. Systems that are used on several complex problems will be the easiest to justify in terms of benefits and costs.

The third level of capabilities extend the ability of groups (and each participant in the group) to explore options, in essence refining judgement. The decision support at this level involves judgement support for multi-grained evaluation of what is being discussed, i.e. fine-grained changes in the comparison of information. The judgement support techniques lay out the details of what options are to be considered more clearly, so that comparisons can be made at fine, medium, and coarse grains of information abstraction. More synthesis as well as detailed exploration should be possible through mutli-grained examination of the analyses produced with level 2 capabilities.

Besides the methods and tools comprising the building blocks of participatory GIS, human personnel and hardware architectures also play an important role. In addition to GIS analysts who process spatial data transactions, PGIS requires also a facilitator(s) who can guide participants through a collaborative decision making process and compensate for their lack of experience with decision support tools. Hardware architectures for PGIS can range from simple local area networks to wide area networks supporting a community-based participation distributed in space and time. New technologies such as mobile computing and map use powered by wireless communication will in the near future expand architectural arrangements for PGIS and allow for a wider participation, especially collaborative work distributed in space and time.

# References

Armstrong, M.P. and Densham, P.J. (1995) "Cartographic support for collaborative spatial decision making", *Proceedings, Auto-Carto 12*, Bethesda, MD, American Congress on Surveying and Mapping: 49–58.

Bayesian Systems Inc. (2000) Available at: http://www.bayesian.com (accessed March 21, 2000).

Bhargava, H.K., Krishnan, R. and Whinston, A.D. (1994) "On integrating collaboration and decision analysis techniques", *Journal of Organizational Computing*, 4(3): 297–316.

Black, D. (1958) *The Theory of Committee and Elections*, Cambridge, Cambridge University Press.

Bostrom, R.P., Watson, R.T. and Kinney, S.T. (1992) *Computer Augmented Teamwork*, New York, Van Nostrand Reinhold.

Brams, S.J. and Fishburn. P.C. (1991) "Alternative voting systems", in L.S. Maisel (ed.) *Political Parties and Elections in the United States: An Encyclopedia*, vol. 1, New York, Garland: 23–31.

Carver, S.J. (1991) "Integrating multi-criteria evaluation with geographical information systems", *International Journal of Geographical Information Systems*, 5(3): 321–339.

Chun, K.J. and Park, H.K. (1998) "Examining the conflicting results of GDSS research", *Information and Management*, 33: 313–25.

Conklin, J. and Begeman, M.L. (1989). "gIBIS: a tool for all reasons", *Journal for the American Society of Information Science*, 40(3): 200–13.

Danish Hydraulic Institute (2000) MIKE11. Available at: http://www.dhi.dk/mike11/index.htm (accessed March 21, 2000).

Densham, P.J. (1991) "Spatial decision support systems", in D.J. Maguire, M.F. Goodchild and D.W. Rhind (eds) *Geographical Information Systems: Principles and Applications*, New York, John Wiley & Sons: 403–12.

Dewey, J. (1933) *How we Think, a Restatement of the Relation of Reflective Thinking to the Educative Process*, Boston, DC, Heath and Company.

Dummett, M. (1997) *Principles of Electoral Systems*, Oxford, Oxford University Press.

Eastman, J.R., Jin, W., Kyem, P.A.K. and Toledano, J. (1995) "Raster procedures for multi-criteria/multi objective decisions", *Photogrammetric Engineering and Remote Sensing*, 61(5): 539–47.

Electrical Power Research Institute (1998) Available at: http://www.smartplaces.com (accessed March 21, 2000).

ESRI (1995) ArcView 2.1, Redlands, CA. Environmental Systems Research Institute, Inc.

Expert Choice Inc. (1994) Expert Choice Reference Manual. Expert Choice, Inc., Pittsburgh, PA.

Expert Choice Inc. (1995) *Team Expert Choice Reference Manual*. Expert Choice, Inc. Pittsburgh, PA.

Faber, B., Wallace, W. and Cuthbertson, J. (1995) "Advances in collaborative GIS for land resource negotiation", *Proceedings, GIS '95*, Ninth Annual Symposium on Geographic Information Systems, Vancouver, BC, March 1995, GIS World, Inc., Vol. 1: 183–9.

Faber, B.G., Wallace, W.W. and Miller, R.M.P. (1996) "Collaborative modeling for environmental decision making", *Proceedings*, GIS '96 Symposium, Vancouver, BC, March. CD-ROM, no pages.

Gehlrein, W.V., Gopinath, B, Lagarias, J.C. and Fishburn, P.C. (1982) "Optimal pairs of score vectors for positional scoring rules", *Applied Mathematics and Optimization*, 8: 302–24.

GeoChoice Inc. (2000) Available at http://www.geochoice.com (accessed March 21, 2000).

GroupSystems (2000) Available at: www.groupsystems.com (accessed March 21, 2000).

Hwang, C. and Lin, M. (1987) *Group Decision Making Under Multiple Criteria*, Lecture Notes in Economics and Mathematical Systems 281, New York, Springer-Verlag.

Jankowski, P., Andrienko, G. and Andrienko, N. (2001) "Map-centered exploratory approach to multiple criteria spatial decision making", *International Journal of Geographical Information Science*, 15(2).

Jankowski, P., Lotov, A. and Gusev, D. (1999) "Application of a Multiple Criteria Trade-off Approach to Spatial Decision Making", in J.C. Thill (ed.) *Spatial Multicriteria Decision Making and Analysis*, Aldershot, Ashgate: 127–48.

Jankowski, P., Nyerges, T.L., Smith, A., Moore, T.J. and Horvath, E. (1997) "Spatial group choice: a SDSS tool for collaborative spatial decision making", *International Journal of Geographical Information Science*, 11(6): 577–602.

Jankowski, P. (1995) "Integrating geographic information systems and multiple criteria decision making methods", *International Journal of Geographical Information Systems*, 9(3): 251–73.

Jankowski, P. and Stasik, M. (1997) "Spatial understanding and decision support system: a prototype for public GIS", *Transactions in GIS*, 2(1): 73–84.

Janssen, R. and Rietveld, P. (1990) "Multicriteria analysis and geographical information systems: an application to agricultural land use in the Netherlands", in H.J. Scholten and J.C.H. Stillwell (eds) *Geographical Information Systems for Urban and Regional Planning*, Dordrecht, Kluwer: 129–39.

Jarke, M. (1986) "Knowledge sharing and negotiation support in multiperson decision support systems", *Decision Support Systems*, 2: 93–102.

Jelassi, M.T., Jarke, M. and Stohr, E.A. (1985) "Designing a generalized multiple criteria decision support system", *Journal of Management Information Systems*, 1(4): 24–43.

Jiang, H. and Eastman, J.R. (2000) "Application of fuzzy measures in multi-criteria evaluation in GIS", *International Journal of Geographical Information Science*, 14(2): 173–84.

Kanafani, A., Khattak, A. and Dahlgren, J. (1994) "A planning methodology for intelligent urban transportation systems", *Transportation Research — C*, 2(4): 197–215.

Lake, R.W. (1993) "Planning and applied geography: positivism, ethics, and geographic information systems", *Progress in Human Geography*, 17(3): 404–13.

Lewis, F. (1994) *A Brief Introduction to Group Support Systems*, Bellingham, WA, MeetingWorks Associates http://www.entsol.com (also available from Enterprise Solutions, Seattle, WA).

Lotov, A., Bourmistrova, L., Efremov, R., Bushenkov, V., Buber, A. and Brainin, N. (2000) "Screening of strategies for water quality planning", submitted to *Journal of Hydroinformatics*.

Malczewski, J. (1996) "A GIS-based approach to multiple criteria group decision making", *International Journal of Geographical Information Systems*, v. 10: 955–71.

Nyerges, T. (1993) "How do people use geographical information systems?", in Medyckyj-Scott, D. and Hearnshaw, H. (eds) *Human Factors in Geographical Information Systems*, London, Belhaven Press: 37–50.

Nyerges, T.L. and Chrisman, N.R. (1994) Facilitating Collaborative Learning and Critical Thinking in a Geographic Information System Laboratory for Undergraduate Education, funds awarded by National Science Foundation, Division of Undergraduate Education, Instrumentation and Laboratory Improvement Program, DUE-9450934.

Nyerges, T., Montejano, R., Oshiro, C. and Dadswell, M. (1998) "Group-based geographic information systems for transportation site selection", *Transportation Research C: Emerging Technologies*, 5(6): 349–69.

Oravec, J. (1996) *Virtual Individuals, Virtual Groups*, Cambridge, UK, Press Syndicate.

Ozernoy, V.M. (1991) "Choosing the best multiple criteria decision-making method", *INFO*, 30(2): 159–71.

Pawlak, Z. and Slowinski, R. (1994) "Decision analysis using rough sets", *International Transactions in Operational Research*, 1(1): 107–14.

Pickles, J. (1996) "Tool or science? GIS, technoscience, and the theoretical turn", *Annals of the Association of American Geographers*, 87(2): 363–72.

Quinlan, J.R. (1993) *C4.5: Programs for Machine Learning*, San Mateo, Morgan Kaufmann Publishers.

Reitsma, R., Zigurs, I., Lewis, C., Wilson, V. and Sloane, A. (1996) "Experiments with simulation models in water-resources negotiation", *Journal of Water Resources Planning and Management*, 122(1): 64–70.

Reitsma, R. (1996) "Structure and support of water-resources management and decision making", *Journal of Hydrology*, 177: 253–68.

Renn, O., Webler, T., Rakel, H., Dienel, P. and Johnson, B. (1993) "Public participation in decision making: a three-step procedure", *Policy Sciences*, 26: 189–214.

Rumbaugh, J., Jacobson, I. and Booch, G. (1999) *The Unified Modeling Language Reference Manual*, Reading, Mass, Addison-Wesley Longman.

Schwartz, R.M. (1994) *The Skilled Facilitator*, San Francisco, Jossey-Bass.

Sheppard, E. (1995) "GIS and society: towards a research agenda", *Cartography and Geographic Information Systems*, 22(1): 251–317.

Spencer, L. (1989) *Winning Through Participation*, Dubuque, IA, Kendall/Hunt Publishing.

Simon, H. (1977) *The New Science of Management Decision*, 3rd edn, Englewood Cliffs, NJ, Prentice-Hall.

Simon, H. (1979) "Rational decision making in business organizations", *American Economic Review*, 69: 493–513.

Stasik, M. (1999) *Collaborative Planning and Decision Making under Distributed Space and Time Conditions*, unpublished doctoral dissertation, University of Idaho.

Stern, P.C. and Fineberg, H.V. (eds) (1996) *Understanding Risk: Informing Decisions in a Democratic Society*, Washington, DC, National Academy Press.

Trochim, W. (1989) "An introduction to concept mapping for planning and evaluation", *Evaluation and Program Planning*, 12(1): 1–16.

Vogel, D.R. (1993) "Electronic meeting support", *The Environmental Professional*, 15(2): 198–206.

von Winterfeldt, D. (1987) "Value tree analysis: an introduction and an application to offshore oil drilling", in P.R. Kleindorfer and H.C. Kunreuther (eds) *Insuring and Managing Hazardous Risks: From Seveso to Bhopal and Beyond*, Berlin, Springer-Verlag: 349–85.

Voogd, H. (1983) *Multicriteria Evaluation For Urban And Regional Planning*, London, Pion.

Webler, T., Levine, D., Rakel, H. and Renn, O. (1991) "A novel approach to reducing the group delphi", *Technological Forecasting and Social Change*, 39: 253–63.

Zeiler, M. (1999) *Modeling Our World: The ESRI Guide to Geodatabase Design*, ESRI Press.

# 4   Social-behavioral research strategies for investigating the use of participatory geographic information systems

## Abstract

Social-behavioral research about the use of participatory geographic information systems requires an informed balance among three research domains — substantive, theoretical, and methodological — if we are to make balanced progress in participatory geographic information science. In this chapter, material for discussing a substantive domain draws from the past few years of the co-authors' research on GIS-supported collaborative decision making, and for the theoretical domain we draw from our development of Enhanced Adaptive Structuration Theory. Out of our research experience in the methodological domain, we develop a new framework for understanding choices among research strategies for social-behavioral studies of participatory geographic information systems use. A research strategy is comprised of several phases: research problem articulation, treatment mode selection, data gathering strategy, data analysis strategy, and reporting strategy. Planning a research study is a matter of making choices within those phases of a research strategy. Informed choices can be made based on criteria relating to the quality of findings we can anticipate. The criteria include strategic considerations for research finding outcomes, as well as validity and reliability. A reinterpretation of internal validity in terms of correspondences among relations with research domains is presented. Several research strategies and their respective phase choices are compared against each other. This systematic treatment of strategies helps researchers understand the advantages and disadvantages of choosing various strategies for studying group use of participatory geographic information systems.

As we strive to understand the social-behavioral implications of advanced information technologies, it can be said that studying the use of GIS is as important as developing the technology itself. Unfortunately, anecdotal evidence about the substantive uses of GIS, although good for sharing experiences and telling stories, has not been sufficient to understand the complexity of how GIS technology is intertwined with social, economic, power and similar structures. Without systematic knowledge of group use of GIS based on social-behavioral research, poor technology designs are likely to be reproduced again and again having (sometimes unintended) social implications for efficiency, effectiveness and equity in group settings (Pickles 1995, 1997). Without a systematic approach to researching GIS use, stories and experiences are difficult to integrate, hence we are less likely to accrue a "knowledge about use", although knowledge accrual is the basis of all science, including geographic information science and a subfield to which we contribute called participatory geographic information science.

Our systematic approach for research methods proceeds as follows. In the next section we address our rationale for what we believe is a contribution toward a more robust methodological turn in participatory geographic information science than published elsewhere. As part of the framework for that turn, the second section adopts Brinberg and McGrath's (1985) perspective for empirical, social-behavioral research process. That perspective consists of three stages of research (planning, implementation, and corroboration), whereby each stage involves a balance among three research domains: substance, theory and method. We describe how the balance among the domains establishes a research orientation for a research study. As the last contribution in the section, we use the three stages of research as a context for articulating six generic phases in any research strategy for social-behavioral research; these phases are research problem articulation, treatment mode selection, data gathering strategy (composed of data setting and data collection strategies), data analysis strategy, and reporting strategy. In the third section, we use each of those research strategy phases to elucidate characteristics about the options and trade-offs in each respective phase. However, we emphasize that the choices by phase are not independent of other phases, and no single strategy works for all research. In the fourth section we present a subset of the research strategies, and discuss how certain tactics can be formulated to avoid threats to the quality of research findings. We use our research as a basis for describing the research strategies and associated tactics. The chapter concludes with a discussion about the balance among research domains that facilitate a methodological turn that has supported development of the material reported in Chapters 5, 6, and 7, and more generally, our recommendations for further progress in participatory geographic information science.

## 4.1 Toward a methodological turn in participatory geographic information science

The focus of this chapter is on formulating social-behavioral research strategies for studying group use of PGIS. We attempt to clarify terminology about *research strategy*, and a portion of a strategy called *research design*, by synthesizing concepts from a number of perspectives in the social-behavioral science literature, particularly from researchers who have treated information use and technology in group contexts (Brinberg and McGrath 1985, Cook and Campbell 1979, Kidder and Judd 1986, Runkel and McGrath 1972, Yin 1994, Zigurs 1993). The synthesis results in a comprehensive, normative characterization of a research strategy composed of several phases: research question articulation, treatment mode selection, data gathering strategy, data analysis strategy, and reporting strategy. Our approach to research strategy incorporates Runkel and McGrath's (1972) view of a research strategy as "study setting" (part of what we have called data gathering strategy), but goes beyond that and combines it with Yin's (1994) view of a research strategy as a series of stages. The synthesis we have created organizes material in such a way that it is now easier to formulate a research strategy by considering options and trade-offs in terms of criteria that characterize "potential qualities in research findings". To describe the quality of social-behavioral research findings we combine the material from Brinberg and McGrath's (1985) non-traditional view of "validity as correspondence" together with Yin's (1994) presentation of the more traditional view of internal, external, and construct validity and reliability. Like the researchers just mentioned, we recognize that the potential quality of research findings are threatened by various contingencies throughout the research process. Recognizing the threats and incorporating tactics to deal with them early on in the research process should help researchers address the threats more effectively.

Contributing to knowledge about the threats to the quality of research findings is part of making an "empirical, methodological turn" along with the contribution to "theoretical turn" in Chapter 2. Such turns help elucidate a change of perspective from GIS viewed simply as a technology (tool), to GIS viewed as geographic information science (Goodchild *et al.* 1999, Mark 2000, Sheppard *et al.* 1999), perhaps at the opposite end of a continuum to tools (Wright, Goodchild and Proctor 1997a, 1997b), to GISystems and GIScience viewed not so much as polar opposites on a continuum, but as an intimate relationship between concept and tool (Fisher 1998). Making such turns help us take a step toward developing a "participatory geographic information science" as a subfield contribution to geographic information science.

A GIS (and hence PGIS) tool development perspective is rather different from a geographic information science perspective, but only at the extremes. Seldom does anyone adopt a perspective of extremes for actual research. A relationship between the perspectives fosters a contribution to each (Fisher 1998). That relationship is elucidated through research that incorporates tool development, tool use, and GIScience as presented in this book. Emphasis on research about GIS tool development, GIS tool use, and concepts developed as part of GIScience can be understood in terms of a balance among three research domains: substance, theory, and methodology. Brinberg and McGrath (1985) elucidate how a balance among those domains motivates, directs, and eventually provides an opportunity for social-behavioral research finding development, and subsequently knowledge building. Such knowledge building where methodology matters significantly is characteristic of research about human-computer interaction (McGrath 1995). Recognizing a balance among the three domains helps one to understand the character of alternative research strategies for social-behavioral research in general, and studies about PGIS use in particular. Understanding the balance among domains and the implications for research strategies can help synergize research about GIS tool development (e.g. Densham, Armstrong and Kemp 1995) and research about social-behavioral implications of tool use (Pickles 1995, Sheppard 1995), as they reconstruct each other as suggested in Chapter 2. A deeper understanding of this issue fosters a synthesis between the seemingly polar perspectives of tool development, i.e. constructionist, and critical social theory, i.e. deconstructionist, research (Miller 1995). Such a synthesis fosters a reconstructivist perspective, accommodating both tool development and tool critique simultaneously (Orlikowski 1992) through concept change. In promoting a reconstructivist perspective, we move beyond a substance versus theory versus methodology debate to show how each research domain is in some way dependent on the other. Each domain is pursued at the expense of the others, permitting us to come away with a new sense of research possibilities.

Although we discuss all three research domains in this chapter, our contribution is intended to be about choosing among research strategies, and the phases that compose them. In other words, we pursue a methodological approach for investigating group use of PGIS from a social-behavioral science perspective. The material in this chapter has broad implications for all studies about the use of advanced information technology, and even social-behavioral science research in general, given the way the material is developed. Nonetheless, our focus will remain social-behavioral studies about PGIS use.

## 4.2 Stages and domains of empirical, social-behavioral research about PGIS use

Research is a process during which there are many pitfalls and traps that direct and constrain the character of empirical findings. In this chapter we adopt a view presented by Brinberg and McGrath (1985) that a research process (particularly an empirical one) consists generally of three stages:

1    planning the research, as in writing a research plan (sometimes submitted in a proposal);

2   doing the research, as in implementing a plan (sometimes more or less successfully); and
3   making a case to corroborate the results (by comparing findings one against another).

Although fewer or more stages can be argued — and certainly this is something of a simplification — these three stages describe research as a process of having an idea of what you want to do, following through on that idea, and then assessing whether the findings make sense within a body of literature by comparing related findings.

Planning a research study is a matter of making choices within phases of a research strategy. The phases are research problem articulation, treatment mode selection, data gathering strategy, data analysis strategy, and reporting strategy. Clearly, research problem articulation sets a stage for what research is pursued, and we will treat this issue in a general way below as a matter of "research orientation" related to our own research agenda. More importantly, this chapter focuses on choices among characteristics within the phases of a research strategy, and subsequently choices among research strategies overall. Furthermore, choices made with regard to treatment modes influence data gathering strategy, so phases are not independent of each other. Choices made with regard to data gathering strategy are rather significant, since such choices can facilitate and/or constrain data analysis strategy. Choices about analysis strategy lead to the outcomes presented as part of the reporting strategy as an attempt to corroborate findings. It is important to note that all phases of research are equally important in composing a comprehensive research strategy.

To help elucidate choices within research strategy phases, we further borrow from Brinberg and McGrath (1985) who suggest that *research orientation* in a research study is a matter of choosing a balanced emphasis among three research domains — substance, theory, and method — throughout the three (planning, implementation, and corroboration) stages of research. What many researchers fail to realize is that all three domains are involved at all three stages of research, whether we recognize explicitly (fostering high quality research) or not (fostering low quality research). The fundamental difference among research orientations stems from which of the three domains is used to provide lead emphasis in a study, which domain is used to support the lead domain, and which domain is emphasized least, and actually constrains the others. Using that perspective, Brinberg and McGrath (1985) describe the difference in basic, applied, and method-driven research based on a lead domain being theoretical, substantive, or methodological, respectively. Consequently, a balanced emphasis among domains that plays out as *research orientation* dramatically influences the type of research findings in any particular study. In fact, problem articulation occurs as a matter of elucidating an

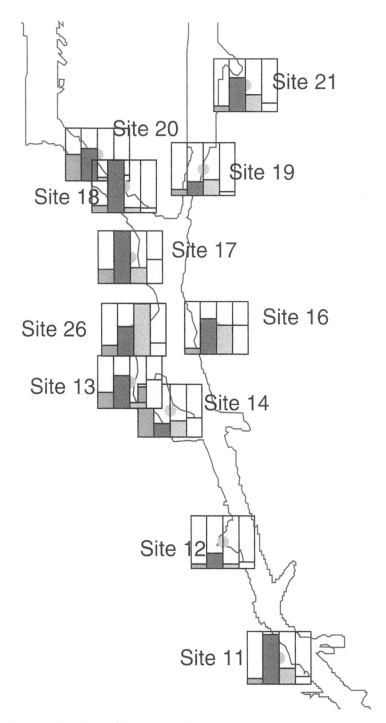

*Plate 1* A histobar display of habitat attribute data in GeoChoicePerspectives

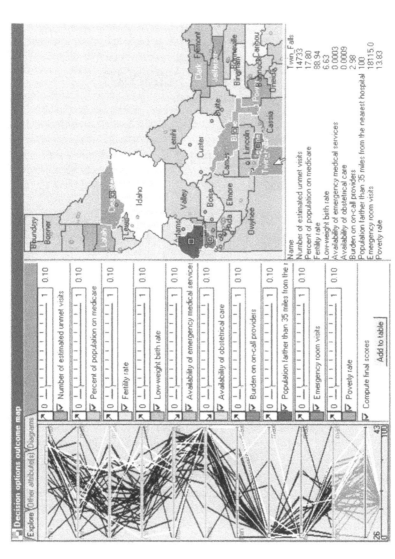

*Plate 2* Value path/map allows a simultaneous exploration/visualization of decision options (Idaho counties in this case), both on the map and on the parallel coordinates graph

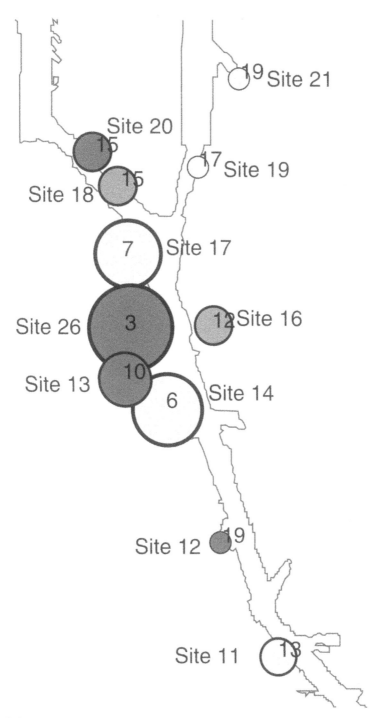

*Plate 3* Consensus rank map of habitat site options

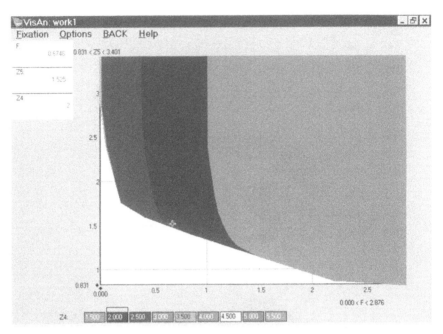

*Plate 4* Decision map representing trade-offs among: water treatment cost — F (in $ billion), concentration of oil pollutants — Z5 (in relative units) and nitrates concentration — Z4 (in relative units)

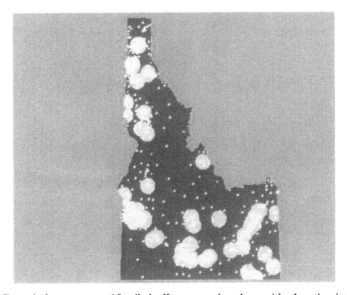

*Plate 5* Gray circles represent 15-mile buffers around each provider location in Idaho. Yellow points are block group centroids representing population locations. The demand for primary health care services is calculated as the function of the population number in each block group, population age and gender. About 10% of demand is located outside the 15-mile buffers and represents the population with inadequate access to primary health care services

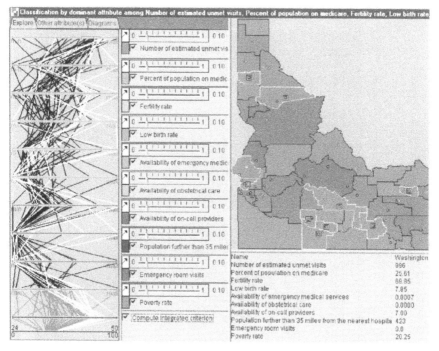

*Plate 6* Decision options outcome map promotes a simultaneous viewing of the performance of decision options on parallel coordinates graph and on the map

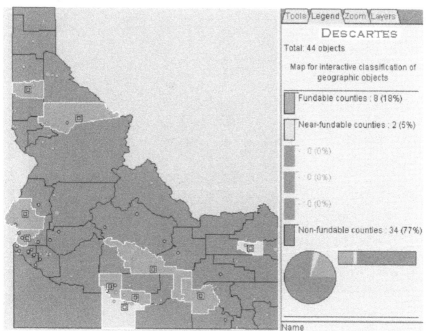

*Plate 7* Interactive classification map with three user-specified decision option classes: fundable, near-fundable, and non-fundable

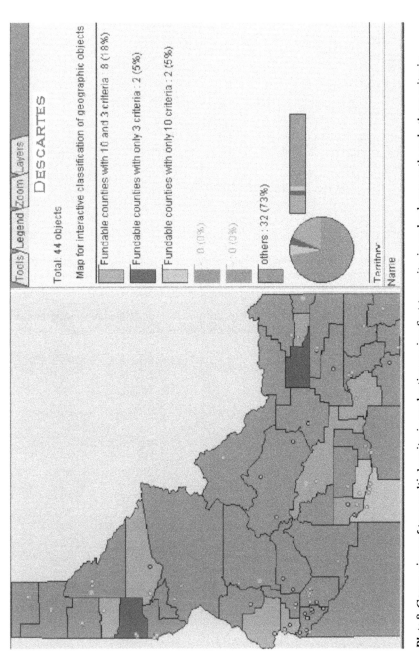

*Plate 8* Comparison of two multiple criteria evaluations using first ten criteria and subsequently only three criteria

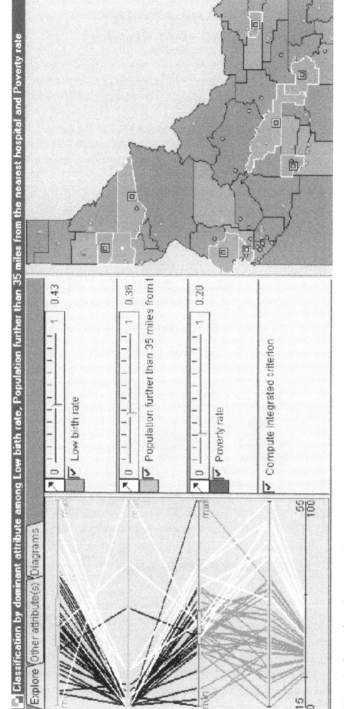

*Plate 9* Noticeable change in the criterion weights resulted in only minor changes in the order of top scoring counties signifying the low sensitivity of evaluation criteria

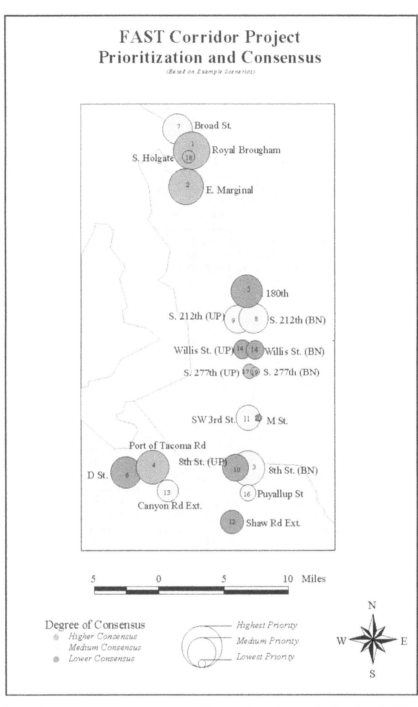

**FAST Corridor Project
Prioritization and Consensus**
*(Based on Example Scenarios)*

7  Broad St.

1  Royal Brougham
S. Holgate  18

2  E. Marginal

5  180th

S. 212th (UP)  9  8  S. 212th (BN)

Willis St. (UP) 14 14 Willis St. (BN)

S. 277th (UP) 17 19 S. 277th (BN)

SW 3rd St.  11  M St.

Port of Tacoma Rd
8th St. (UP)
D St.  6  4  10  3  8th St. (BN)

13  16  Puyallup St
Canyon Rd Ext.

12  Shaw Rd Ext.

5      0      5      10  Miles

Degree of Consensus
● Higher Consensus          Highest Priority
   Medium Consensus          Medium Priority
● Lower Consensus           Lowest Priority

N
W      E
S

*Plate 10* Consensus map of FAST freight mobility project evaluation in a decision
experiment

interesting topic in terms of the lead domain, as there are theoretical, substantive, or methodological research problems in almost every research area. None of the orientations is really any better than any others overall. They all contribute to empirical science. "Good" research is research that is conducted well. However, as we describe later in this chapter, certain types of research are preferred by certain funding agencies. Thus, the orientation of the research does matter in regards to the type of research findings anticipated.

As examples of research orientation, we consider the three relevant chapters in the next section of this book. Chapter 5 is representative of an applied research orientation using substance, method, concept. We lead with a substantive decision problem concerning primary care for rural health in the State of Idaho, followed by a methodological domain using task analysis to describe the tools used during the decision process and, third, informed by the conceptual domain of Enhanced Adaptive Structuration Theory 2 (EAST2) to check on the significance of the constructs treated. Chapter 6 is representative of an applied research orientation using substance, concept, method. We lead with a substantive topic about transportation improvement program decision making. We structure this in terms of the conceptual domain by using EAST to inform us about the constructs and premises to treat. We use those to implement a proposition analysis as a case study method directed by the premises. In Chapter 7, we use a basic research orientation using concept, method, and substance. We lead with EAST as a conceptual understanding and implement an experimental design that was informed by a habitat site selection problem for the participants to perform. So far we have not used a method-driven orientation because we are not performing research on social-behavioral methods.

We bring to the reader's attention that "GIS methods" about which we research are actually part of the substantive domain, rather than the methodological domain. This is what situates our research in a context of geographic information science. Furthermore, although a lead domain sets the emphasis (hence the orientation) of a research study, we should remember that all three domains are important. Without a contribution from each of the three domains to some sufficient level in that balance, a study (or research proposal for that matter) could be considered incomplete, or at best only a partial study (proposal), left open to criticism, and thus more easily dismissed.

In detailing the three research domains, like Brinberg and McGrath (1985), we recognize three levels (here we call them abstraction levels) for treating what is at issue in each of the domains (see Table 4.1 for an example). The levels of abstraction are *elements*, *relations*, and *embedding context* (embedding context was referred to by Brinberg and McGrath as "embedding systems"). Each research domain has a set of elements, relations, and embedding context relevant to that domain. However, a

*Table 4.1* Levels and domains for characterizing social-behavioral research: an example from collaborative decision making about habitat redevelopment

| Abstraction levels | Research domains | | |
|---|---|---|---|
| | substantive | conceptual (theoretical) | methodological |
| elements (e.g. actors, sites and information technology) | participants from organizations, fish and wildlife habitat redevelopment projects as land parcel locations (sites and situations) along the Duwamish Waterway as viewed by stakeholders (organizations, groups, etc.); geographic information technology they use to make decisions | categories of potential sites and the actors interested in promoting sites; human values for land described in terms of the constructs and aspects within | variables and associated treatment modes for characterizing elements (actors, sites, technology) of substantive domain in line with conceptual domain; e.g. types of maps and decision tables being used, number and experience of decision actors |
| relations (e.g. human behaviors using information technology as a group) | decision actors and potential habitat projects in relation to each other; social/power relations among decision makers; projects as alternative sites and situations for redevelopment | growth and decline of habitat and its impact on the local scene; actor interaction in terms of premises of EAST2 (see Figure 2.1) | social-behavioral analysis of the participants interacting in a decision situation using elements (variables in line with associated treatments) mentioned above |
| embedding context (e.g. land area and inter-organizational context) | Seattle urban area context as setting for salmon habitat redevelopment; governmental instruments for addressing habitat loss; private–public perspectives for geographic information technology use | human-environment significance of habitat based on interpretations of what is valued; plus the ontological and epistemological context for EAST2 | social-behavioral research strategy to address research questions; research strategy phases include treatment strategy, data gathering strategy, analysis strategy and reporting strategy |

fundamental activity in research is drawing a correspondence among the elements of the domains, and in essence the respective relations from each of the domains (Brinberg and McGrath 1985). For example, in Table 4.1 the participants and their relationship to information technology (maps and decision tables as information structures) in the substantive are conceptualized as decision actors with certain perspectives about those information structures in the conceptual domain. In turn, those decision actors and their impressions of information structures need be measured as "coded" variables in the methodological domain in order to provide data for relationships that are to be analyzed. Drawing a correspondence among the relations, hence respective elements, of the three domains are the basis of generating "valid" empirical outcomes in the second stage of research, i.e. doing research, particularly data analysis (Brinberg and McGrath 1985). However, we believe that the *planned* character of the correspondence, i.e. setting up the potential for correspondence in the first stage as a research plan, is just as important. Re-interpreting the correspondence in the third stage, i.e. the verification stage, is also just as important. Consequently, it is best to "plan" what elements/relationships in the substantive domain must correspond with what details/relations are used to describe them in the conceptual domain, and those both must correspond to methodological variables/procedures that we use to collect data, if we are to "plan on" attaining quality findings in a study. Such a plan documents the research strategy as a fundamental part of a proposal. How that plan is actually carried out in the course of "doing the research" might be different, but a well thought out plan (core of the proposal) enhances the possibility of effective plan implementation, i.e. carrying out the research.

Based on a perspective about balanced emphasis among domains, in the subsections to follow we treat each of the levels of abstraction for each of the substantive, theoretical and methodological domains to set up a discussion about evaluation and choice among research strategies. That sequence of presentation of the domains emulates the orientation of our long-term research agenda, and not necessarily any particular study concerning research related to use of PGIS that we have used. In addition, it highlights the reason for including this chapter about the significance of choosing among research strategies.

### 4.2.1 Substantive domain: situation as topic, place, and time

The substantive domain is perhaps the most open-ended, as well as "specific" of all three domains. It is open-ended in a sense that there is a tremendous opportunity to examine group use of PGIS. It is specific in the sense that "topic, place, and time matters" — what happens in one place may or may not be similar to another time and/or place. We need a substantive context (or focus) to ground the character of theory and methodology. So, we overview the substance of our own research to provide context for what comes later in the Chapter. In our research we have been concerned with public–private problems related to land use, environmental cleanup, salmon habitat redevelopment, transportation planning, and health planning in the Pacific Northwest, USA. All of these are rather long-term concerns, in that the concerns have been with us for some time and are likely to continue to influence our everyday lives for some time to come.

Two of the most significant substantive *elements* are problem-topic and people. In land use planning we dealt with land use scenarios for Latah County in Idaho, whereby students as interested and affected parties configured a land use plan (Jankowski and Stasik 1997a, 1997b, Stasik 1999). For environmental cleanup we have been examining decisions about remediation efforts at the Hanford, Washington Reservation and how decision makers, technical specialists and interested and affected parties take part in the process (Drew *et al.* 2000). We have examined the concerns about habitat redevelopment in the Duwamish Waterway in Seattle, Washington (Jankowski *et al.* 1997, Nyerges *et al.* 1998a, Jankowski and Nyerges 2001). In transportation planning we have been dealing with transportation improvement decision making for the central Puget Sound area as concerns local, regional, and state decision analysts (Nyerges *et al.* 1998b). For health planning we have been dealing with resource allocation for primary health care at the county level in the State of Idaho (Chapter 5). In all cases we have adopted realistic decision situations as the focus of our substantive domain. All of the participants have used GIS directly or indirectly. The level of technology has differed across the various decision situations. Some of the GIS technology has been commercial off-the-shelf, whereas other GIS technology has been prototype experimental. We have examined both local area network-based (Jankowski *et al.* 1997, Nyerges *et al.* 1998b) and internet-based approaches to participatory use of GIS (Jankowski and Stasik 1997a, 1997b).

*Relations* among the people and the technology can be understood as the day-to-day communication among the people, as well as their disposition toward GIS technology use. To a large degree we have been interested in understanding the character of the human-computer-human interaction that occurs in decision situations on a project by project basis, i.e. from one project to another of both similar and different kinds. We are interested in the dynamics of decision processes, and how GIS technology plays a role in those processes relative to the types of problems we are studying.

The *embedding context* for the public–private problems in the Pacific Northwest, USA is the place and time situation for participatory activities of an inter-organizational character. The local culture of place is very much a part of the embedded contextual relations in the substantive domain. It has been reported in local newspapers that the City of Seattle is the "public participation" capital of the USA. Whether this is accurate or not does not matter as much as that "place and time" are ripe for studying group use of GIS within realistic situations. Organizations and their bureaucracies together with social norms help structure the planning situations in the Pacific Northwest. Given the nature of the problems of our interest, we are very much aware that inter-organizational decision making cuts across a variety of scales and organizations. It is important to recognize that a multitude of socio-political relationships exist among people in groups, groups in organizations, organizations in communities, and communities in society, whereby actors at any level can act as agents of change.

### 4.2.2  Conceptual domain: Enhanced Adaptive Structuration Theory 2

Rather than duplicate what has been presented in Chapter 2 in full, here we provide a brief overview of the significance of the theoretical domain. As has been mentioned previously, in the theoretical domain, choice of a theory (or building a theory) for articulating what to expect during human-computer-human interaction provides a way of "systematically" interpreting how people make use of GIS in a problem context. A theory by its nature is a systematic abstraction of the elements and relations of the substantive domain. Thus, we use EAST2 in the hope of coming to a better understanding of the elements, relations, and embedding contexts described above for the substantive domain.

EAST2 consists of a set of eight constructs detailed in terms of 25 aspects (the *elements* of the conceptual domain) that describe significant issues for characterizing group decision making (see "conceptual domain" in Table 4.1). The seven premises of EAST2 describe the *relations* among the eight constructs, hence relations among the 25 aspects. The structuration process of what/who influences what/who is the *embedding context* for EAST2. Neither technological nor social character of an organization predominates in change — they work together to structure, and hence reconstruct each other — the fundamental idea underlying "adaptive structuration".

Following from the respective premises of Table 2.5 in Chapter 2, the example research questions of Table 4.2 have been articulated with reference to a public–private decision situation involving habitat redevelopment in the Duwamish Waterway of Seattle Washington. These questions are related to "geographical worlds constructed by society" and/or the "use of geographical concepts to think about geographical phenomena and make decisions about places" as topics of long-term significance suggested by Goodchild *et al.* (1999, p. 738). The premises of Table 4.2 are general in form, such that the reader could develop research questions to suit their own topical situation related to use of PGIS, perhaps something like those in the table. Whatever questions are used to motivate the research, nouns

*Table 4.2* Research questions motivated by premises of EAST2: an example from collaborative decision making about habitat redevelopment

| Premises | Research question motivated by respective premise |
|---|---|
| *Convening premises* | |
| **Premise 1.** Social-institutional influences affect the appropriation of group participant influences and/or social–technical influences | – How does decision task complexity influence the types of geographic information structures, e.g. maps, tables, diagrams, appropriated by the participants? |
| **Premise 2.** Group participant influences affect the appropriation of social-institutional influences and/or social-technical influences | – Does prior knowledge/experience with decision aids promote more use of maps and decision models?<br>– Does knowledge acquired through group decision making participation promote more effective use of decision models? |
| **Premise 3.** Social-technical influences affect the appropriation of social-institutional influences and/or group participant influences | – What is the relationship between the usage of maps appropriated in relation to decision models?<br>– What kinds of decision models are appropriated in relation to maps?<br>– Does technology setting have any influence on the type of decision aid use? |
| *Process premises* | |
| **Premise 4.** Appropriation of influences affect the dynamics of social interaction described in terms of group processes | – What is the relationship between maps usage and decision functions?<br>– What is the relationship between decision table usage and decision functions?<br>– Are there differences in the level of group conflict associated with different decision phases?<br>– Does task complexity influence appropriation during a given phase?<br>– Does group conflict have any influence on appropriation of decision aids? |
| **Premise 5.** Group processes have an affect on the types of influences that emerge during those processes, and emergent influences affect the appropriation of influences | – What new kinds of information structures are called for as a result of group interaction at the inter-organizational level?<br>– What new information emerges as a result of changes in task management? |

*Table 4.2* Continued

| Premises | Research question motivated by respective premise |
|---|---|
| *Outcome premises* | |
| **Premise 6.** Given particular influences being appropriated, if successful appropriation occurs and group processes fit the task, then desired outcomes result | – Given that a group appropriates a particular type of information structure that has been found to be useful in the past, and if the information structure is appropriated during "specific conditions", can we expect the outcome from the process to be satisfactory to all participants? <br> – What structure appropriation under what conditions of group process appears to affect the dependence of the decision outcome on group participant structuring? |
| **Premise 7.** Given particular influences being appropriated, if successful appropriation occurs and group processes fit the task, then reproduction of social-institutional influences result | – In what way are various social-institutional structures together with group participant structures linked to the opportunity to challenge the task outcome? <br> – How do inter-organizational protocols and the social interaction during group process promote or discourage further group work? |

that detail constructs in the questions inevitably translate into methodo-logical variables for which data are collected (or not) using a choice of research strategies as described in the following sections.

The seven premises of EAST2, together with the respective example questions, indicate that a wide variety of interesting, empirical research opportunities exist. A significant point to be made here is that EAST2 addresses geographic information technology, a substantive problem, and group participation all as part of the substantive domain to be investi-gated. As such, EAST2 is a start at "taking the theoretical turn" in geo-graphic information science that emphasizes the social-behavioral implications of GIS use (Pickles 1997). Addressing any one or more ques-tions among such a wide variety of questions is a considerable challenge. The challenge "begs" for a systematic approach for evaluation of empirical social-behavioral research strategies so that we can better understand how empirical findings relate to each other in our attempts to build knowledge about the implications of PGIS use. In a social-behavioral research study, choices about treatments, data gathering strategies, data analysis strategies and reporting strategies all follow from research question articulation, hence we consider these issues next.

### 4.2.3 *Methodological domain: a social-behavioral perspective*

We agree with McGrath (1995) that "methodology matters" for studies about human-computer interaction, and in particular those that investigate human-computer-human interaction in our own research. Methodology matters in a significant way for social-behavioral research studies about PGIS because choices in method, called here "research strategy", can lead to different qualities for empirical findings. Just how and why these choices matter, and the influence such choices have on quality of findings, is the focus of this chapter. A reader will find that our interest in this topic about making choices in research strategy is itself a "decision strategy" similar to what was discussed in Chapter 2. McGrath (1982) made similar claims about the importance of making choices, but cautioned that such choices could not be set up as a "lock step" system of rules. Nonetheless, we believe that options and trade-offs can be elucidated so that researchers can make informed choices.

Despite the large number of textbooks on research methods, there appears to be no clear comprehensive framework for understanding how various types of research strategies compare and contrast to one another. Runkel and McGrath (1972) were among the first to present a framework for comparing research strategies in what they called a "research strategy circumplex" consisting of eight strategies. However, tracing the lineage back from McGrath (1995) shows that there are misinterpretations and gaps in the framework. Oddly, Brinberg and McGrath (1985) treat concerns about validity throughout the research process, but they do not relate that validity to research strategy as presented by McGrath (1982), which was an elaboration of the Runkel and McGrath (1972) material. From an independent but related perspective, Yin (1993, 1994) provides a comprehensive treatment of case study methodology, but compares ethnography, sample surveys, and historical analysis to case studies in only a brief manner. Others, such as Williams, Rice and Rogers (1988) for so-called new, communications technology media and Zigurs (1993) for group support systems research, provide long lists of research approaches, but fall short of comparing one to another to provide deeper insight about why one strategy might be more useful than another for a particular situation. Still others, such as Scarbrough and Tanenbaum (1998), make reference to research strategies for social science research in terms of the analytical basis in such strategies, rather than all aspects of research method. In this chapter we synthesize much of that work, re-examining the nature of various components of research strategy, and elucidating frameworks for those components to show that methodology, and in particular research strategy formulation, is a matter of making choices. Like McGrath (1982) cautioned, we cannot set out one specific set of rules for what is the most appropriate overall because there are so many contingencies; however, we can say that the choices are now clearer for a variety of research strategies.

Methodology, defined in the *Oxford English Dictionary* (OED) as the theory and science of methods, should inform us about how to work with empirical data as we address research questions about PGIS use. Methodology should help us make difficult choices about research strategy so we can address our research questions in the best way we can, and arrive at research findings that are of high quality. The term "method" according to the OED means: "A special form of procedure adopted in any branch of mental activity, whether for the purpose of teaching and exposition, or for that of investigation and inquiry", that is the sense of the term we adopt here. Method qualified by either "qualitative" or "quantitative" will not help us in this discussion. Using such adjectives is not useful, beyond a rather general categorization that soon becomes much more complex, hence these terms are avoided here in favor of more descriptive terms for methods. What we are after here is an attempt to understand better the quality of research designs (as the core of research strategies) from a broad perspective, hence research strategies, as they provide a means to attain a potential quality in research findings. As such, we are looking for relative comparisons among phases of research designs as a matter of informed choice with implications. Trustworthiness, credibility, confirmability, dependability, reliability, and validity are six criteria for expressing quality of research designs (Brinberg and McGrath 1985, Einsenhart and Borko 1993, Kidder and Judd 1986, Yin 1994). Validity and reliability appear to be two criteria that cover the four others well enough, given the different ways validity and reliability have been expressed in the literature.

Based on McGrath's circumplex of research strategies, we re-examined the basis of research methods applicable for human-computer-human interaction research and devised an alternative and broader, but at the same time, more detailed perspective than the strategy circumplex. This results in a more foundational attempt to "sort through the choices" for research strategies. We must caution, however, that like decision strategies, research strategies are but recommended approaches, e.g. as in an agenda. What research (strategy) actually transpires in the process of "doing research" might be similar to or entirely different from what is planned, depending upon opportunities and circumstances encountered when a researcher actually moves forward. Nonetheless, it is possible to elucidate research strategies in a normative manner, some of which we have used, showing the similarities and differences in approaches, as we remember that "methodology matters".

Starting with the notion of research strategy, we synthesize the idea of stages of research (Brinberg and McGrath 1985) together with phases of research (Yin 1994) as presented in Table 4.3. This synthesis results in three stages with six phases in each stage. Our use of the concept research strategy composed of phases is more like that of Yin (1994), but the details of the phases as we present in following sections are grounded in the work of Runkel and McGrath (1972) and Brinberg and McGrath (1985). Within a stage of research we see six normative phases concerning: research questions, treatment modes, study setting, data collection (the former two together called a data gathering strategy), data analysis, and reporting of

*Table 4.3* Stages of research and phases in a research strategy

| | |
|---|---|
| **Stage 1: scoping/planning the research** | |
| Phase 1) scope/consider research questions | |
| Phase 2) scope/consider treatment mode(s) strategy | Devising |
| Phase 3) scope/consider setting strategy | research |
| Phase 4) scope/consider data collection strategy | design |
| Phase 5) scope/consider analysis strategy | (phases 2–5) |
| Phase 6) scope/consider reporting strategy | |

| | |
|---|---|
| **Stage 2: doing/implementing the research strategy** | |
| Phase 1) articulate/commit research questions | |
| Phase 2) specify/commit treatment mode(s) strategy | Implementing |
| Phase 3) specify/commit setting strategy | research |
| Phase 4) specify/perform data collection strategy | design |
| Phase 5) specify/perform analysis strategy | (phases 2–5) |
| Phase 6) perform reporting strategy | |

**Stage 3: corroborating the research findings**
Compare to other findings and compare/contrast the validities;
take each phase 1–6 above and report connections to other findings in term of
similarities, differences, limitations, biases, etc.

findings. Our synthesis sheds light on opportunities for explicit choices, whether researchers chose to make them or not.

Although one would hope to move through stages of research and phases of a research strategy in a linear manner, that seldom occurs because of unknowns and/or contingencies. One of the biggest challenges in research is to understand the trade-offs in research design (see Table 4.3). The core of research design comes with establishing treatment modes (phase 2). However, establishing such modes has significant implications for data gathering strategy (phases 3 and 4) and data analysis strategy (phase 5), and vice versa. Thus, we include all four phases as part of research design. We briefly describe each of these phases to provide an overview of research strategy options.

For any given study, research questions can be motivated by concerns in any one or more of the three research domains: substantive, conceptual, and methodological. One of those domains commonly becomes the "lead domain", that is, the perspective through which the questions will be addressed. Brinberg and McGrath (1985) describe the lead domain as the domain that provides the "orientation" of the research. If the questions are substantive in character then the lead domain is the substantive domain. If the questions are theoretically motivated then the lead domain becomes the conceptual domain. If the questions are methodological in character then the lead domain is the methodological domain. However, all three domains are part of the research activity, so research orientation is a matter of which domain leads, which domain supports, and which domain constrains.

The *elements* of the methodological domain are called variables. The variables within research questions of concern are treated in the next phase in terms of what are called treatment modes. Selecting one or more treatment modes for the variables expressed in terms of units of observation should follow from the nature of the questions being asked. Treatment modes establish relationships among units of observation, some or all are measured while some or others are controlled.

Data gathering strategy enables the implementation of treatment mode through the collection of data in a given setting. A data gathering strategy is a method for collecting data as enabled/constrained by researcher choices of the setting in which the participants find themselves; thus, a data collection instrument is used to record the observations from the setting. Certain combinations of data collection instruments and study settings create a circumstance that enables and/or prohibits certain kinds of data being made available for analysis. As such the data gathering strategy establishes the *embedding context* of the data. Later in this chapter we present various objectives for data gathering strategies which can be used to compare one strategy to another; this in turn helps us to understand the implications for research quality from different perspectives.

Once having established a data gathering strategy then it is important to consider data analysis strategy. The analysis strategy is a method that attempts to examine *relations* between and among observations. Those relations are a reflection of what is sought in a well articulated research question. Data analysis strategy follows from treatment mode specification and is constrained by data gathering strategy, taking into consideration the measurement levels of the data by mode. Data analysis strategy is composed of a selection of procedures commonly called techniques, although some researchers prefer to call them methods. Analysis techniques enable a set of relations to be examined which should cross-match to the other two domains. When the techniques of the methodological domain do not implement the same relational comparisons as desired in the conceptual and substantive domains, then the quality of findings is jeopardized.

The last phase of the research strategy is the reporting strategy. A reporting strategy specifies how to report specific findings. Some basic issues about reports are the schedule of reports, i.e. interim reports, final report; intended audience; what kinds of information structures are going to be used to appeal to the audience; and the expected distribution of the findings. An important part of reporting strategy is the link a researcher sets up for establishing the opportunity to corroborate findings. A researcher should want an audience to be aware of the limitations and bias of results, before being criticized and discredited for not making them known.

The "normative" phases of a research strategy described above provide a framework for listing a set of research strategy options together with their characteristic components (see Table 4.4). A more comprehensive

*Table 4.4* Characteristic components of research strategy phases

| Research strategy | Forms of research questions addressed | Treatment modes | Data gathering (setting and data collection) strategy | Data analysis strategy | Reporting strategy |
|---|---|---|---|---|---|
| laboratory experiment (Yin 1994, McGrath 1995) | how,why | K: topic X: tasks and sessions Y: event code details M: specific event codes I: all other events | laboratory experiment code development | nonparametric and parametric inferential, code analysis – Analysis of Variance (ANOVA) – regression – Exploratory Sequential Data Analysis (ESDA) | Human Computer Human Interaction (HCHI) relationships |
| experimental simulation (McGrath 1995) | how, why 2 | K: topic X: tasks and sessions Y: event code details M: specific event codes I: all other events | experiment simulation | descriptive inferential statistics | interaction among actors with fixed tasks |
| case study (Yin 1994, Zigurs 1993) | how, why | K: timeframe X: tasks Y: event details M: specific event codes | case codings using multiple instruments | content structuring analysis | details of events |
| field experiment (McGrath 1995, Zigurs 1993) | how, why | K: topic X: tasks and sessions Y: event code details M: specific event codes I: all other events | field experiment | – parametric inferential code analysis – nonparametric inferential, sequential code analysis | dynamics of processess and/or outcomes |
| field study (Zigurs 1993, McGrath 1995, Runkel and McGrath 1972) | who, what, where, how many, how much | K: timeframe X: tasks Y: event details M: specific event codes | structured interview | content structuring analysis | character of content and context of a situation |

*Table 4.4* Continued

| Research strategy | Forms of research questions addressed | Treatment modes | Data gathering (setting and data collection) strategy | Data analysis strategy | Reporting strategy |
|---|---|---|---|---|---|
| ethnography (Zigurs 1993, Yin 1994) | who, what where | Y: construct categories Y: relationships | direct observation | interpretative code content analysis | relationships among actors and/or events |
| field survey (Yin 1994, McGrath 1995) | who, what, where, how many, how much | K: topic X: tasks and sessions Y: event code details M: specific event codes I: all other events | survey instrument | parametric inferential statistics – ANOVA – regression | overview of situation |
| archival analysis (Yin 1994) | who, what, where, how many, how much | K: topic X: tasks and sessions Y: event code details M: specific event codes I: all other events | archive acquisition | parametric inferential statistics – regression | overview of trends |
| historical analysis (Yin 1994) | how, why | K: topic Y: events I: all other events | documents | chronological sequencing | sequence of events |
| action research (Zigurs 1993, Avison *et al.* 1999) | who, what, where | Y: event activity I: all other events | participate | participatory interpretation | details of realistic interactions |
| application task description (Zigurs 1993) | who, what, where | Y: event outcomes M: coded outcomes I: all other events | enumerate character of tasks | task analysis interpretation | activities and needs that occur during work process |

*Table 4.4* Continued

| Research strategy | Forms of research questions addressed | Treatment modes | Data gathering (setting and data collection) strategy | Data analysis strategy | Reporting strategy |
|---|---|---|---|---|---|
| meta-analysis (Zigurs 1993) | who, what, where, how many, how much, how, why | Y: capture constructs M: coded from Y I: all other constructs not of concern | code studies by categories | category analysis and logical synthesis | synthesis of findings across several studies |
| software usability (Zigurs 1993) | what, how many, how much, how | K: time-frame X: task or event Y: details of actors, task or sessions M: specific tasks | video tape actions | descriptive statistics and content structuring analysis | user-focused complexity of software capabilities |
| formal theory (McGrath 1995) | how, why | Y: capture constructs M: coded from Y I: all other constructs not of concern | literature review | category analysis and logical synthesis | conceptual framework |
| hardware/ facility engineering (Zigurs 1993) | what, how many, how much, how | K: time X: elements of place Y: elements of design | create design | design review | comments about layout of a facility |
| judgement tasks (McGrath 1995) | what, where, how many, how much | K: topic X: tasks and sessions Y: event code details M: specific event codes I: all other events | judgement task | parametric inferential statistics – ANOVA – regression | frequency counts diagram |

*Table 4.4* Continued

| Research strategy | Forms of research questions addressed | Treatment modes | Data gathering (setting and data collection) strategy | Data analysis strategy | Reporting strategy |
|---|---|---|---|---|---|
| computer simulation (McGrath 1995) | how many, how much | K: topic of scenario X: task event categories Y: changes in events M: scenario event R: outside influences on events I: all other events | generate data | simulation scenario analysis | details of dynamic processes |
| theorem proof (Zigurs 1993) | what, how | Y: constructs of theorem M: similar constructs I: all other constructs | assemble logic | logic analysis | logical connections among constructs |

Key:
A variable is treated in the following ways:
K:    constant (fixed)
X:    partition (controlled)
Y:    observed (measured)
M:    matched by design (secondary control)
R:    randomized by design
I:    ignored by design

list of strategies, addressing all social-behavior science research, is beyond the scope of this book. However, the list we provide is a rather substantial contribution in comparison to what other textbooks and monographs have provided. An observant reader at this time might notice that the labels for research strategies in the left-hand column are similar to labels for various components of a strategy, e.g. a data gathering strategy or a data analysis strategy, as found in the literature. The labels for research strategy commonly take their lead from either the data gathering strategy or the data analysis strategy, but seldom a combination of both. It is for that reason that *terminology in research methods are so confusing, and hence difficult to understand without some apprenticeship. Although we also struggle with confusion in terminology in order to sort through it, we hope this table will promote a clearer understanding of a wide array of choices, as well as* the similarities and differences among commonly used terms presented by phase in a strategy. The list of strategies in Table 4.4 attempt to sequence "advantageous combinations" of components from phase to phase. Selecting among strategies is a matter of understanding the advantages and disadvantages of each phase, as each has inherent opportunities and limitations that influence detailed characteristic components among the phases in order to address research questions in the best way possible. We describe the detailed components for each phase of the strategies (listed in Table 4.4) in the following section in order to highlight the trade-offs among the strategies.

Research strategy composition would be straightforward if we were able to establish the detailed character of each phase independently of the other phases. If that were the case we speculate that the degree of chaos and confusion in terminology would not abound and fewer books on research methods would have been written to date. The dilemma is that there are interaction effects that occur among (specific components of) the phases. Hence, some combinations of characteristic components from phase to phase work better together than others. It is for this reason that we have attempted to elucidate the term "research strategy" as a series of phases, although we fully recognize that such phases might evolve as inter-active and circular activities based on incremental understandings of the trade-offs. To sort through this problem we now consider each of the research strategy phases and the characteristics associated with each in order better to elucidate trade-offs.

## 4.3 Comparing components of a research strategy by phase

As described previously, a research strategy is composed of the following: articulating research questions, selecting treatments to establish a basis for analysis, selecting a research setting and data collection techniques that combine to specify a data gathering strategy, analyzing the data in line with the treatments, and finally reporting the findings. Each of these phases will now be addressed in turn as to how it influences choices among research strategy.

### 4.3.1 Articulating research questions as the basis of a research problem

Research questions often ground a research study, whether they are implicit or explicit. Yin (1994) provides a list of five research strategies and the kinds of questions that are most appropriate to address with such strategies. His contribution was to list strategies and enumerate research questions that each strategy might be likely to address. In Table 4.5 we include those five strategies and add 13 other strategies enumerated from several literatures dealing with social-behavioral science research methods (Brinberg and McGrath 1985, McGrath 1995, Rice and Rogers 1988,

*Table 4.5* Strategic considerations for research strategies relevant to studies of PGIS use

| Research strategy | Form of research question (1 for every question) | Objective A generalizable to actors, space, time 1 = no 3 = some 5 = yes | Objective B represent behavioral detail 1 = poor 3 = medium 5 = good | Objective C real context preserved 1 = no 3 = some 5 = yes | Overall score |
|---|---|---|---|---|---|
| meta-analysis (Zigurs 1993) | who, what, where, how many, how much, how, why 7 | 3 | 3 | 1 | 14 |
| field study (Zigurs 1993, McGrath 1995, Runkel and McGrath 1972) | who, what where, how many, how much 5 | 3 | 1 | 5 | 14 |
| laboratory experiment (Yin 1994, McGrath 1995) | how, why 2 | 5 | 5 | 1 | 13 |
| experimental simulation (McGrath 1995) | how, why 2 | 3 | 5 | 3 | 13 |
| ethnography (Zigurs 1993) | who, what, where 3 | 3 | 1 | 5 | 12 |
| field survey (Yin 1994, McGrath 1995) | who, what, where, how many, how much 5 | 5 | 1 | 1 | 12 |
| archival analysis (Yin 1994) | who, what, where, how many, how much 5 | 3 | 1 | 3 | 12 |
| case study (Yin 1994, Zigurs 1993) | how, why 2 | 3 | 1 | 5 | 11 |
| field experiment (McGrath 1995, Zigurs 1993) | how, why 2 | 3 | 3 | 3 | 11 |

*Table 4.5* Continued

| Research strategy | forms of research question (1 for every question) | Objective A generalizable to actors, space, time 1 = no 3 = some 5 = yes | Objective B represent behavioral detail 1 = poor 3 = medium 5 = good | Objective C real context preserved 1 = no 3 = some 5 = yes | Overall score |
|---|---|---|---|---|---|
| usability test (Zigurs 1993) | what, how many, how much, how 4 | 1 | 3 | 3 | 11 |
| action research (Zigurs 1993, Avison *et al.* 1999) | who, what where 3 | 1 | 1 | 5 | 10 |
| application description (Zigurs 1993) | who, what, where 3 | 3 | 1 | 3 | 10 |
| judgement tasks (McGrath 1995) | what, where, how many, how much, 5 | 3 | 1 | 1 | 10 |
| historical analysis (Yin 1994) | how, why 2 | 3 | 1 | 3 | 9 |
| hardware/ facility engineering (Zigurs 1993) | what, how many, how much, how 4 | 1 | 1 | 3 | 9 |
| formal theory (McGrath 1995) | how, why 2 | 3 | 1 | 1 | 7 |
| computer simulation (McGrath 1995) | how many, how much 2 | 1 | 1 | 1 | 5 |
| theorem proof (Zigurs 1993) | what, how 2 | 1 | 1 | 1 | 5 |

Runkel and McGrath 1972, Scarbrough and Tanenbaum 1998, Yin 1993, 1994, Zigurs 1993). These particular strategies have been selected because we believe they have potential for studies about PGIS use. Because there are more strategies than types of questions of concern, we can expect that multiple strategies address the same (similar) kinds of questions.

Research questions lead us to anticipate potential research findings as one type of strategic consideration in choosing a research strategy.

Brinberg and McGrath (1985) list three strategic considerations for research strategies; here we call them objectives that establish the methodological context for research findings:

a  generalizability to actors, space, time;
b  representation of behavioral detail; and
c  preservation of real context.

As such, these strategic considerations become the objectives of a research strategy for investigating potential findings. No single research strategy can optimize all three objectives simultaneously. Therefore, all research strategies have some inherent shortcomings if we have an interest in all three objectives. The challenge is to make the trade-offs with regard to those objectives.

We have scored the research strategies with regard to potential to address research questions and ability to excel at addressing an objective. With regard to questions, each of the questions (what, when, where, etc.) has been given one point in the scoring, meaning that the more questions a strategy can address, the more valuable a commitment of energy might be to that strategy for a particular study, and perhaps over the long-term as well. The objectives have been scored by identifying the best outcome on potential findings and assigning between one and five points as appropriate. Consequently, the ranks within the cells in Table 4.5 are an indicator of how well that research strategy might address the various questions and objectives. The strategy scores in the right-hand column of the table are an overall indicator of comprehensiveness of the strategy relative to each other, implying a "breadth" of the strategy for addressing questions and strategic considerations. In our own research we have been interested in strategies that address questions of "what and how", as well as strategies that address "how and why" for human-computer-human interaction in PGIS use. The application description scores slightly lower, whereas the case study and experimental research strategies score rather higher. Nonetheless, each has served its purpose as described later in this chapter and in Chapters 5, 6 and 7, respectively. The reader can make their choices of strategies based on interest in the trade-offs in the types of questions as well as the strategic objective considerations.

If choosing a research strategy was simply a matter of evaluating the considerations for the research question phase, then the task would be easy. What we actually have in Table 4.5 is the beginning (research questions) and end (potential characteristics of research findings). However, we need to address at least the phases for treatment mode, data gathering strategy (composed of research study setting and data collection) and data analysis strategy in order to be truly informed about choices. All of those phases relate to research design. We now turn to treatment mode as a core concern in research design.

### *4.3.2 Treatment mode*

Research questions have embedded within them the terms needed to identify variables for data collection and subsequent analysis. Assigning a treatment mode to a variable qualifies the variable in such a way as to establish a potential basis for data analysis. Variables can be assigned to one of six treatment modes: fixed, partition controlled, matched controlled, measured, randomized, or ignored (Runkel and McGrath 1972). A "fixed" variable establishes data for a specific unit of observation that are not to change in relation to the other variables. "Partitioned controlled" variables means that two or more categories for data have been established for units of observation, and the categories are mutually exclusive. "Matched controlled" variables are those for which two or more categories for a unit of observation have been established and the categories overlap to "match observations" among the categories. "Measured" variables are those whereby data are sampled from a freely changing set of potential observations — free at least in regards to an absence of researcher influence as much as possible. "Randomized variables" are those for which data being sampled are equally at chance to be observed according to that unit as in any similar situation. "Ignored variables" are those for which a researcher knows that a variable exists, but chooses intentionally not to do anything with the units of observations. Again, an illustration using the same example would be useful. Whenever a variable is not fixed, partition controlled, matched controlled, measured, or randomized, then it is ignored by implication. Unintentionally ignoring variables sets up a greater chance of introducing confounding variables into an analysis and subsequent interpretation.

As an example of treatment modes, in the Duwamish habitat laboratory experiment (reported in Chapter 7) we used the following treatment modes. We "fixed" the setting to be a Department of Geography decision lab with access to Spatial Group Choice software. Group participation was "partitioned controlled" to be five-person groups with the same membership throughout the experiment. We applied "match control" to the five decision tasks to differentiate decision situations. This match control established the exposure of participants to the number of sites (8 or 20), the number of criteria (3 or 11), the combinations of which created tasks 1–4. The simplest task (i.e. task 1) involved eight sites described by three criteria, and the most complex (i.e. task 4) involved 20 sites each described by 11 criteria. In each of the four tasks, participants had individual access to Spatial Group Choice software. In task 5, participants were exposed to a task with 20 sites and 11 criteria; matching task 4, however, participants had group-as-a-whole access to Spatial Group Choice software, thus differentiating task 4 from task 5 in terms of access to technology. For our dependent variables we "observed (i.e. measured through use of videotape coding)" three types of data: decision aid structure appropriation to represent information use, decision functions to be used in determining decision phases, and group working relations to code social conflict among particip- ants. We "randomized" assignment of participants to a group (hence expertise with habitat/geographic information that might have influenced outcomes). Finally, we ignored gender bias/participation in groups, time of day on influence of interaction, and probably also a number of other vari- ables. Supposedly, the advantage of laboratory experiments is that they are the most highly controlled of research strategies when it comes to treatment modes.

The variables of the methodological domain are the elements that correspond (cross-match) to the phenomena of the substantive domain and the theoretical aspects of the conceptual domain. Researchers select variables and their corresponding treatments based on prior substantive and/or theoretical knowledge and interest in relationships among ele- ments. Assignment of treatment mode in the methodological domain is a way of establishing relationships between and among data to be used in analysis. Thus, the core of a research design is based on selecting these treatment modes, whether this be an exploratory research, e.g. using field study or ethnography research strategy, in which most variables are meas- ured, or in a laboratory experiment, in which controlled treatments are paramount (see Table 4.6). Assigning a treatment mode provides a way of specifying relations in the methodological domain in line with the substan- tive relations, which are guided by relations from the conceptual domain. If the relations in the conceptual domain are unknown, perhaps due to the study being exploratory, then further interpretation from a variety of per- spectives in the substantive domain are necessary. A major issue about assigning treatment modes concerns construct validity. Since construct validity refers to making operational the measures for the conceptual ele- ments (Yin 1994), and hence the representations of the substantive ele- ments, as well to make sure our measurements represent what we purport, we must be vigilant of the way we implement variables as a basis for relations (Brinberg and McGrath 1985). If independent and dependent variables are to be used in analysis, these can actually follow from the treatments established. However, we note that the labels "inde- pendent/dependent" are guided by the relations in the conceptual domain, rather than the methodological domain. Causality is a matter of concep- tual and not methodological underpinnings. Nonetheless, assigning treat- ment mode is the principal way of establishing correspondence among the elements of the three domains.

Different research strategies can support different treatment modes due to the influence of data gathering strategies commonly associated with those research strategies. Table 4.6 provides an idea of the potential use of treatment modes within the 18 research strategies from Table 4.4. As can be seen, all research strategies have at least one mode, that being "Y:measured". Even in the most open-ended, exploratory studies, all

*Table 4.6* Potential use of treatment modes for selected research strategies

| Research strategy | Potential use of treatment modes in research strategies | | | | | |
|---|---|---|---|---|---|---|
| | *K:* constant (fixed) | *X:* partition (controlled) by | *Y:* observed (measured) | *M:* matched by design – as a second control | *R:* randomized by design | *Z:* ignored by design |
| hardware/ facility engineering | Yes | Engineer | Maybe | No | No | Yes |
| usability test | Yes | Researcher | Yes | Yes | Yes | Yes |
| judgement task | Yes | Researcher | Yes | Yes | Yes | Yes |
| field survey | Yes | Researcher | Yes | Yes | No | Yes |
| laboratory experiment | Yes | Researcher | Yes | Yes | Yes | Yes |
| archival analysis | Yes | Researcher | Yes | Yes | Maybe | Yes |
| computer simulation | Yes | Researcher | Yes | Yes | Yes | Yes |
| experimental simulation | Yes | Researcher | Yes | Yes | Yes | Yes |
| ethnography | No | No | Yes | Yes | No | Yes |
| field study | No | Coded from Y | Yes | Coded from Y | Maybe | Yes |
| case study | Yes | Coded from Y | Yes | Coded from Y | Yes | Yes |
| historical analysis | Yes | Coded from Y | Yes | Coded from Y | No | Yes |
| action research | No | Participant | Yes | No | No | No |
| field experiment | Yes | Participant/ researcher | Yes | Yes | Maybe | Maybe |

Can a research strategy make use of that mode?
Yes:       the mode is supported
No:        the mode is not supported
Maybe:    the mode might just be supported
Other term: mode is specific to condition described

observations are considered "measured observations". The more the potential treatments available in a given research strategy, the more flexibility there is likely to be to capture the data as motivated by research questions. That is, some research questions cannot be addressed when limited modes are available. The research strategies in the table are ordered on the "X: control" column. This column was chosen because of our interest in wanting to understand how we might establish mode assignments for which strategy was most powerful for questions dealing with "how" and "why", since these questions are the fundamental ones in the premises of EAST2. Other researchers might choose a different ordering based on the importance of research questions enumerated for each strategy listed in Tables 4.4 and 4.5.

### 4.3.3 Data gathering strategy

To make a variable operational, it must have a *unit of observation*. In research about PGIS use, it is important to differentiate the units of observation in terms of *units of interaction* versus *units of organization*. A unit of interaction is the sampling unit of social interaction in terms of the activity of groups, whereas the unit of organization is the grouping of decision actors. Sanderson and Fisher (1994), working in the context of human-computer interaction, clearly point out that units of interaction range from cognitive acts of individual computer users, speech acts of individuals, speech acts directing group attention in meetings, meeting phases of groups, entire meeting sequences of a group project, through to studying a set of group projects over several years. Pasquero (1991) describes how different an inter-organizational level of interaction is from an organizational level of interaction. To a large degree the interaction seems to differ based on the nature of "information flow bureaucracy", a key factor in facilitating (hindering) information flow among participants. No matter whether we work with units of interaction or organization, units of observation for variables later become units of analysis in line with treatment mode assignment; which is why treatment mode consideration precedes data gathering.

Opportunities for data gathering are many, as there are many research settings and many data collection techniques that can be used as part of the *embedding context* of a study. These opportunities for observation are what Brinberg and McGrath (1985) called research strategies, but we prefer to call such opportunities "*data gathering strategies*". Research strategy, as we have described it above, is seen to be a more inclusive term that includes problem articulation, treatment mode selection, data gathering strategy, data analysis strategy, and reporting strategy (Yin 1994). Consequently, we clarify here how a data gathering strategy as embedding context sets the stage for a data analysis strategy. Furthermore, treatment modes addressed in the previous section indicate how to treat variables during data collection. Thus, choice of treatment mode helps guide what data gathering strategy we should choose, since data gathering strategy constrains us to certain kinds of treatment of variables. Implementations of treatments are actually constrained and/or facilitated by data gathering strategy — a circumstance of phase interaction as mentioned previously during the discussion of research strategy.

A data gathering strategy is composed of a research study setting and one or more data collection techniques that combine to facilitate data gathering/collection. Together, a study setting and (set of) technique(s) establish an opportunity for collecting data that meet the needs of treatment mode assignment, following research question articulation.

A study setting for empirical research takes on the character of one of three fundamental types: laboratory, field, and the cross between them called a field laboratory. We would be remiss if we did not recognize that there are important micro variations in each, e.g. an engineering laboratory and a social-behavioral laboratory are meant to structure social-behavioral relations among participants in different ways, with the former more controlled for software usability testing than the latter for social-behavioral studies. Field settings are as varied as the number of places of work, hence some settings are more structured than are others. However, our main point here is that three settings can characterize sufficiently the variation in "researcher induced control over social-behavioral relations" in the setting — the underlying dimension of "research setting" as shown on the left side of Figure 4.1. What this means is that researchers structure lab settings. Participants at work, rather than researchers, socially structure field settings. Researchers and participants together contribute to structuring the social behavioral relations that take place in field laboratories. A researcher must choose among these settings, or chose all three, for any particular study.

Working on research about organizational culture, Sackmann (1991) described the differences in several data collection techniques in terms of a researcher's interest in "pre-structuring the data categories" at the time of data observation. Using the underlying dimension of pre-structuring, we can array data collection techniques from most to least pre-structuring as roughly: survey instrument, document coding, structured interview, group discussion, in-depth interview, and direct observation, as depicted along the bottom of Figure 4.1. Pre-structuring occurs based on a researcher's pre-existing conceptual framework about a topic, i.e. a structuring of variables (data concepts) for which to collect data that a researcher is "locked into". Therefore, "least pre-structuring" can be thought of as *an opportunity for post-structuring* of the data observations through "recoding", since sufficiently rich "discursive or narrative" detail is available.

As examples of pre-structuring in data collection, both closed-ended and open-ended survey instruments pose directed questions devised from a researcher's pre-existing conceptual framework. A document code is

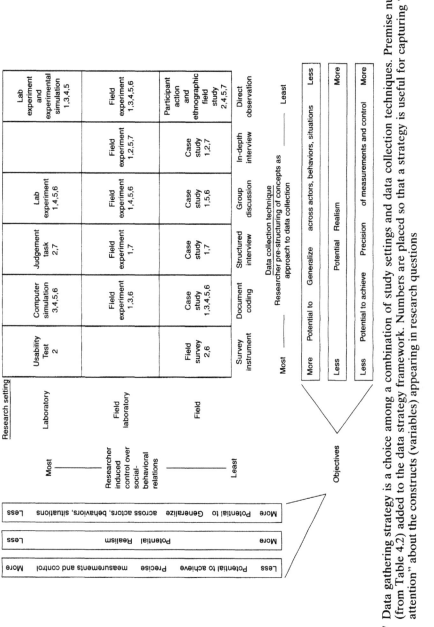

*Figure 4.1* Data gathering strategy is a choice among a combination of study settings and data collection techniques. Premise numbers (from Table 4.2) added to the data strategy framework. Numbers are placed so that a strategy is useful for capturing "group attention" about the constructs (variables) appearing in research questions

usually assigned from a protocol of codes that develops (or at least is refined) while working with the document materials. Structured interviews have a direction in mind, but the interviewee has a chance to direct the conversation. The direction of group discussions is at the whim of the group. An interviewer in an in-depth interview usually lets a subject talk at will — although of course subtopic questions are prepared in advance. In direct observation, e.g. using a video camera, little or no pre-structuring is involved, as one observes whatever the camera captures. Of course, all of the data collection techniques have some amount of pre-structure, but the amount is relatively different along the continuum. A researcher must chose among these techniques, or chose all of them, for any particular study.

By combining the opportunities to choose a study setting with the opportunities to select data collection techniques, we create a data gathering strategy as a choice among an array of data gathering opportunities (see Figure 4.1). We make use of settings, whereby behavior may or may not be influenced. We also make use of pre/post-structuring of data collection techniques, whereby researchers structure data concepts previous to data collection as in a survey instrument, or after data collection as in coding a videotape. In Figure 4.1, the "least pre-structuring" is meant to imply "post-structuring".

The bi-dimensionality of the framework sets up a choice among at least 18 different data gathering strategies, as a choice among different opportunities for data observations. The two dimensions in the framework help us understand trade-offs among approaches to data observations. At the same time, however, the dimensions establish a set of constraints on those trade-offs. All 18 cells are different one from another due to the dimensionality, but clearly they are related as well through neighborliness in position with respect to dimensionality. The cells of the framework can be used to differentiate some of the more common strategies: field survey, case study, computer simulation, judgement tasks, laboratory experiment, experimental simulations, field study, field experiment, and participant action design (see Runkel and McGrath 1972 for details of field survey, judgement task, computer simulation, laboratory experiment, experimental simulations, and field study; see Zmud, Olson, and Hauser 1989 for field experiment; see Onsrud, Pinto and Azad 1992, and Yin 1994 for case study; see Avison *et al.* 1999 for participant action design). None of the strategies in general is better than any other; but given a particular setting and a particular technique, a different set of data are collected for a particular study. Each strategy can contribute to a study in a different way, and it is up to a researcher to make an informed choice about which one or more to pursue for a study. All require resources and time. The framework can also help researchers who are interested in mixed-method approaches to research. Selection of complementary, rather than just supplemental, data strategies for triangulation promotes a more thorough examination of a topic (McGrath 1982). When triangulating in a study, a researcher draws from (or makes use of) at least three different perspectives (or approaches) to interpret findings, each supplementing the interpretation.

We can glean from Figure 4.1 that strategies inherently differ according to a researcher's induced control of social-behavioral relations among participants and a researcher's pre-structuring of the data collection constructs. Those dimensions inform us about possible trade-offs concerning empirical results that are likely. The trade-offs link with the dimensions for empirical findings as presented earlier.

The "objectives" for the potential findings of a study motivate a choice for one or more data gathering strategies. The three objectives are inherently in conflict. There is no way that a researcher can "have it all", i.e. full generalizability, full realism, and full precision of behavioral detail, simultaneously. When you strive to maximize one objective, you must give up on maximizing another. If you want a study to apply to many places and times, you must forego some amount of realism and precision, holding resources constant for a particular study. For example, a field survey can "potentially" provide data that are the most generalizable along each of the fundamental dimensions: induced control of relations and pre-structuring of data concepts. However, field surveys provide little situation specific realism because responses are commonly constrained to certain predefined categories by researchers, and because surveys tend to be completed in settings where respondents are removed from the situations in the question.

A further difficulty is that trade-offs work differently in each of the two dimensions. Note that any given strategy in a cell position along both the social-behavioral relations and the pre-structuring of concept dimensions influences each objective in converse ways. For example, a field survey, associated with the most pre-structuring of concepts, encourages less realism because the instrument can be applied easily and systematically across populations. However, when a field survey is implemented in a field setting, it can be more generalized because a researcher influences the social-behavioral relations the least. Such is the nature of the inherent trade-offs associated with objectives for research findings.

Thus, the power in the framework is that it helps us understand advantages and disadvantages of strategies in general, especially since all strategies have failings for various reasons. It can help a researcher understand the nuances in mixed-method (strategy) research. The framework can help prioritize a research agenda, allowing a research team to plan phases of a study, or multiple studies about a particular substantive topic. However, it can only help with a particular research study or set of studies when a researcher combines the substantive domain and the theoretical domain to help with selection of strategy, since it is these latter domains that motivate social-behavioral research questions about PGIS use.

Choosing (and implementing) a data strategy is intimately linked with issues in the theoretical and substantive domains of a study. A researcher should treat all three research domains when making an informed choice. Based on our own research, we have identified the strategies that are most likely to help us pursue empirical research about the use of PGIS, assigning the premises (thus research questions of Table 4.2) to the framework (see Figure 4.1). The assignment of particular premises (research questions) to cells is based on whether a particular social-behavioral relation is important with regard to the way the setting induces control, and whether a particular data collection technique can sample the interaction of "group attention" in that type of setting. Group attention in this case is meant to represent (crudely) a shared understanding of the situation. Notice that the assignment is not one premise to one strategy. When used in this way, the framework is useful for organizing discussions about the pros and cons of data strategies in terms of specific research premises/questions. However, the framework is not meant to "optimize" for a particular strategy or a particular premise. That would be asking for more "rigor" in the framework than it offers at this time. Other researchers might want to assign labels for their research questions in a different way, but the general framework for data strategy will remain the same.

Inspection of Figure 4.1 shows that there is a bias in the premises toward the lower portion of the figure where field survey, case study, focus group, participant action, and ethnographic field strategies occur due to the influence of real settings. In the top portion of the framework, representative of laboratory experiments and simulations, fewer premises appear because laboratory experiments make use of contrived settings, constraining the opportunity to collect a variety of data, but allowing a fine grain data collection. The trade-off between laboratory studies and field studies are field experiments, which attempt to garner the best of both worlds. Field experiments are among the most difficult strategies to implement. Some believe they are not very useful because of the compromise in social-behavioral control (McGrath 1995), while others believe they are among the most valuable strategies (Zmud, Olson and Hauser 1989). Both points of view are probably correct. A strategy is ineffective when implemented poorly, and effective when implemented well. The answer is in the opportunity that presents itself, because of the difficulty of setting up such experiments.

Further inspection of the framework in Figure 4.1 shows that certain of the premises are indeed associated with certain data strategies, while other premises may be addressed with a variety of data strategies. What this really means is that certain social-behavioral constructs are more sensitive to certain data collection techniques, while others are not. Summarizing the information about premises to be addressed by strategies in Figure 4.1 provides us with the ranking information in Table 4.7. When ranking the strategies from the most to least number of premises addressed by a strategy, we see that case studies and field experiments can address all seven premises. This is only true because multiple data collection techniques are required to collect all appropriate data. Note that some of the research strategies near the bottom of the table treat only 1 premise (the PGIS technology premise), and others do not treat any. In the latter case, "0" premises addressed indicates that such strategies address constructs only (rather than premises). Along the bottom of Table 4.7, the fewest number of strategies to address any particular premise is eight, and no premise is addressed by all 18 strategies. This provides further evidence that choices abound, and are likely to make a difference to research findings. A detailed examination of what strategies are appropriate for what premises should consider the trade-offs in generality, realism, and precision with regard to objectives for research findings. This must be accomplished on a research study by research study basis.

In our own studies to date, we have conducted case studies and laboratory experiments. In a case study about public health resource allocation across counties of Idaho we examined the types of GIS capabilities that public health decision makers put to use; however, we did not have data about how such capabilities were used in the decision process to produce certain outcomes (Chapter 5). In a case study about transportation improvement decision making, we used a case study with document coding to examine the potential for use of advanced group-based GIS capabilities, but again fine grain interaction data about the process was not available (Nyerges *et al.* 1998b, also see Chapter 6). In a case study examining risk factors in cleanup decisions about hazardous waste at Hanford, document coding provided data at a level of coarse granularity, but useful for describing the cyclical nature of decision chains (Drew *et al.* 2000). In a Duwamish River habitat site selection study (Jankowski and Nyerges 2001, Nyerges *et al.* 1998a, also see Chapter 7) we made use of a laboratory experiment strategy that provided data about group interaction with maps and decision models collected at a one-minute resolution coded from video tape. The contrived, experimental setting was useful for examining process interaction in the group decision situation. The decision outcomes were less interesting because the task goal was contrived, even though we used a task that emulated an actual habitat restoration decision situation.

During the timeframe of the studies mentioned above, we have proposed three field experiments on brownfields land redevelopment, habitat restoration, and hazardous waste redevelopment. Zmud, Olson and Hauser (1989) warn that field experiment data strategies are more difficult to plan and implement than most strategies, if not all of the other strategies. Our experience with laboratory experiments and field (case) studies shows this indeed to be the case. In planning field experiment research, Zmud, Olson and Hauser (1989) suggest that a major challenge is to find a realistic situation, in which participants do not mind the introduction of new (PGIS) technology to solve a real problem and at the same time allow

*Table 4.7* Premise investigation supported by data gathering strategy

| Data strategy | Premises (P) | | | | | | | |
|---|---|---|---|---|---|---|---|---|
| | 1 | 2 | 3 | 4 | 5 | 6 | 7 | count |
| case study | × | × | × | × | × | × | × | 7 |
| field experiment | × | × | × | × | × | × | × | 7 |
| field study | × | × | × | × | × | × | × | 7 |
| meta-analysis | × | × | × | × | × | × | × | 7 |
| ethnography | × | × | × | × | × | | × | 6 |
| laboratory experiment | × | | × | × | × | × | | 5 |
| participant action | | × | × | × | × | | × | 5 |
| computer simulation | | | × | × | × | × | | 4 |
| experimental simulation | × | | × | × | × | | | 4 |
| historical analysis | × | × | × | | | × | | 4 |
| archival analysis | | × | | | | × | × | 3 |
| judgement task | | × | | | | | × | 2 |
| field survey | | × | | | | × | | 2 |
| hardware/ facility | × | | × | | | | | 2 |
| usability test | | × | | | | | | 1 |
| formal theory | | | | | | | | 0 |
| theorem proof | | | | | | | | 0 |
| application description | | | | | | | | 0 |
| **Strategy count by premise** | 9 | 11 | 11 | 9 | 9 | 9 | 8 | |

researchers to examine the situation; people are said to be distrustful of new technology and fear examination of processes when "real work" is expected. However, we have not had that problem, since groups have been willing to participate. Instead, our experiences with data gathering strategies for field experiments suggest that the core difficulty is the transition from problem articulation to data gathering strategy in the context of a substantive situation. Field experiment strategies must balance the completion of a real task with a modified setting, and at the same time collect meaningful data on convening, process, and outcome aspects of the task. The above mentioned studies, both undertaken and anticipated, encouraged us to believe that certain types of research questions (generalized here as premises) were likely to be best supported by certain data gathering strategies, hence the purpose for systematically exploring that connection in this chapter. Extending that connection from one phase to the next, we now turn to data analysis strategy.

### 4.3.4 Data analysis strategy

Without data collection, there is no data analysis. Consequently, one fundamental issue concerns the kind of units of observation that are available to be used as the units of analysis. The units of analysis are the basic elements of the methodological domain that match the elements of the substantive and conceptual domains in a data analysis. It is common that observation units must be transformed into recoded units for analysis, but the units that are sought are those that represent the elements in relations to be examined. As mentioned previously, analysis is about examining *relations* between elements (variables) in the methodological domain. Relations between variables in the methodological domain take their lead from either relations among phenomena in the substantive domain or relations among concepts (constructs) in the theoretical domain depending on the orientation of study, i.e. relationships that are the foundation of applied or basic research in a particular context, respectively. The treatment modes for variables in the methodological domain are established based on what (or how) variables relate to what variables. Of course, all variables could be measured variables (treatments) as in an exploratory study. Eventually, however, some variables become the foundation for interpreting relations; these are the ones that first become controlled, perhaps as matched-controlled, and not necessarily independent variables which would imply *causality*. In the context of our PGIS research, we first look to the substantive domain for interesting events related to public–private decision making that concern geographic information, and then ask how the theoretical domain informs us about potential causality among the substantive elements, for example, how stakeholders might behave with each other as per premise 2, or in the context of generating maps, as information structures as per premise 3 (in Figure 2.1 and Table

4.2). The relations in the methodological domain are implemented through social-behavioral data analysis procedures for manipulating units of analysis according to units of organization, i.e. actors as groups of participants (see Table 4.8 for a list of data analysis strategies). Data analysis strategies have been selected on the basis that they are the dominant analysis characteristics of the research strategies in Table 4.4, which were selected, as mentioned previously, because of the potential use for studying PGIS use.

The data analysis strategies make use of a varying number of treatment modes. More treatment modes allow a researcher to attain more specific results from analysis. In Table 4.8 it appears that more treatment modes are also associated with higher levels of measurement. With higher levels of measurement a researcher can more easily distinguish differences in elements that take part in social-behavioral relations. Traditionally, a level of measurement (as specified in Table 4.8) has always been an indicator of information detail in an element, but indicates little about the character of the relation. In social-behavioral research, the concept of *internal validity* has been used as a criteria to describe the quality of the potential findings when causal relations are anticipated in the conceptual domain, and whether such relations can be investigated in the methodological domain through the use of analysis procedures (Cook and Campbell 1979, Kidder and Judd 1986, Yin 1994). In rethinking the nature of that potential quality of findings uncovered through analysis, Brinberg and McGrath (1985) used the idea of correspondence among relation *features* to describe the details of the cross-match of relations across the domains. Ten features provide a description of "potential information gain" from an analysis, each feature contributing to information gain when it is deemed a characteristic of the relation. Brinberg and McGrath (1985 p. 96–7) list the following ten features (used in the right-hand column of Table 4.8) intrinsic to such relations.

1   Statements of the *presence or absence* of a relation for any pair of elements $i$ and $j$ can be interpreted to be the presence or absence of a relation between elements $i$ and $j$, for example, whether decision tables and maps are used in relation to each other in a habitat redevelopment decision situation.

2   Statements of the *temporal order* of the pair of elements $i$ and $j$ can include statements that $i$ precedes $j$, and $j$ precedes $i$, or that $i$ and $j$ are simultaneous, for example, whether map use precedes decision table use, or vice versa.

3   Statements of *logical order*, or logical direction, of the relation between $i$ and $j$ can include statements that $i$ leads to $j$, that $j$ leads to $i$, or that $i$ and $j$ each affect each other, for example, decision table use leads to consensus aid use in a habitat redevelopment decision situation.

*Table 4.8* Characteristic components associated with data analysis strategies

| Data analysis strategy (also called "data analysis technique" | Treatment modes* | Levels of measurement for data in variables | Potential feature correspondence based on relation analysis** |
|---|---|---|---|
| content structuring analysis | K: timeframe<br>X: tasks<br>Y: even details<br>M: specific event codes | categorical narrative | 1–3, 8, 9, 10 |
| interpretative code content analysis | Y: construct categories<br>Y: relationships | categorical narrative | 1, 2, 8, 9, 10 |
| simulation scenario analysis | K: topic of scenario<br>X: task event categories<br>Y: changes in events<br>M: scenario event<br>R: outside influences on events<br>I: all other events | interval and ratio | 1–10 |
| category analysis and logical synthesis | Y: capture constructs<br>M: coded from Y<br>I: all other constructs not of concern | categorical and ordinal | 1, 2, 8, 9, 10 |
| descriptive statistics<br>– content<br>– count relationships<br>– frequency counts | X: task or event<br>Y: aspects of actors, tasks or sessions | interval and ratio | 1, 2 |
| non-parametric inferential, sequential code data analysis, e.g. ESDA<br>– lag sequential<br>– cycles<br>– transitions | K: topic<br>X: tasks and sessions<br>Y: event code details<br>M: specific event codes<br>I: all other events | ordinal, interval and ratio | 1–3, 8, 9, 10 |
| parametric inferential code analysis, e.g.<br>– ANOVA<br>– regression | K: topic<br>X: tasks and sessions<br>Y: event code details<br>M: specific event codes<br>I: all other events | interval and ratio | 1–7 |
| chronological sequencing | K: topic<br>Y: events<br>I: all other events | categorical and ordinal | 1, 2 |
| participatory interpretaion | Y: event activity<br>I: all other events | categorical and ordinal | 1, 2, 8, 9, 10 |
| task analysis interpretation | Y: event outcomes<br>M: coded outcomes<br>I: all other events | categorical and ordinal | 1, 2, 3, 8, 9, 10 |

*Table 4.8* Continued

| Data analysis strategy (also called "data analysis technique" | Treatment modes* | Levels of measurement for data in variables | Potential feature correspondence based on relation analysis** |
|---|---|---|---|
| design review | K: time<br>X: elements of place<br>Y: elements of design | categorical | 1 |
| logic analysis | Y: constructs of theorem<br>M: similar constructs<br>I: all other constructs | categorical and ordinal | 1, 2, 3 |

\*   Treatment modes:
K:   constant (fixed)
X:   partition (controlled)
Y:   observed (measured)
M:   matched by design (secondary control)
R:   randomized by design
I:   ignored by design

\*\*   Features of a relation:
1 – presence or absence of a relation
2 – temporal order of a relation
3 – logical order of a relation
4 – direction of a functional relation
5 – form of a functional relation
6 – deterministic or stochastic character of a relation
7 – temporal stability of a relation
8 – element $g$ to element $i$ of the $ij$ relation
9 – element $k$ to element $j$ of the $ij$ relation
10 – element $l$ to both $i$ and $j$ or to the $ij$ relation itself

4   Statements of the *direction of a functional relation* between $i$ and $j$ can include statements that $i$ and $j$ are positively related or inversely related, for example, with an increase in task complexity there is an increase in decision table use.

5   Statements of the *form of a functional relation* between $i$ and $j$ can include statements that $i$ and $j$ are related in a linear (and therefore monotonic) form, or in a nonlinear but monotonic form, or in a non-monotonic but single-peaked form, or in a multiple-peaked form of more complexity, such as a recurring cycle, for example, with increases in the length of time within a decision situation a group experiences more conflict.

6   Statements of the *deterministic or stochastic character* of the $ij$ relation include statements that $i$ and $j$ are related though a fixed-functional or probabilistic means from one event stage to the next, for example, groups are more likely to use a public display screen than they are individual display screens in the context of *idea integration* = "bringing ideas together" rather *idea differentiation* = "pursuing separate paths of work".

7 Statements of the *temporal stability* of the *ij* relation can include statements that the *ij* relation is stable over time, or that it is stable but with minor fluctuations, or that it changes in a stable (regular) pattern, or that it changes in a variable pattern, for example, map use is used in problem exploration more often than are decision tables.

The seven statements refer to features intrinsic to relations between any two elements *i* and *j*. They have representations in each of the three domains that are different but similar, because the elements in each domain are different but similar. Because each relation is set within a context of relations, several additional features of any *ij* relation involve features extrinsic to that *ij* pair but intrinsic to the set of overall relations within a context. That is, the extrinsic features tie each *ij* relation to other elements and relations within a context of relations under consideration. These extrinsic features of *ij* relations include the following three features.

8 Statements of the relation of each other *element* g *to element* i of the *ij* relation. These elements *g* represent antecedents of the *ij* relation. These *gi* relations are to be assessed on the seven intrinsic features of relations listed above, for example, an external element such as hardware contributing to the failure of the Spatial (ArcView) software module (within the Spatial Group Choice software) that is related to map use rather than to decision table use could be considered on the seven features above.

9 Statements of the relation of each other *element* k *to element* j of the *ij* relation. These elements *k* represent alternative antecedents (alternative to *i*) of the *ij* relation. These *kj* relations, also, are to be assessed on the seven intrinsic features of relations listed above, for example, an external element, such as hardware failure related to the Choice module supporting decision table use within the Spatial Group Choice software, could be considered on the seven features above.

10 Statements of the relation of each other *element* l *to both* i *and* j or to the *ij* relation itself. These elements *l* represent moderator variables that may alter, influence, mediate or modulate the *ij* relation. These *l-ij* relations, also, are to be assessed on the seven intrinsic features of relations listed above, for example, an external element, such as failure of the Group Choice software, related to both map use and decision table use, could be considered on the seven features above.

The extrinsic features 8, 9, 10 contain all of the "extra-relation" effects on *i*, *j*, and *ij*. These are the places in the research process for many of the influences and/or effects, anticipated and otherwise, that arise in research that cannot be accounted for, i.e. confounding influences and/or effects. As such, they are places that threaten validity of findings.

Each of the above features can be viewed as a test that contributes to building information about a relation, and hence contributes to building a sense of the validity (quality) about what we know about the relation between the elements. Furthermore, what is interesting about the ten features is that they "unpack" the character of "internal validity", so that we can now understand a relation as a range from "existence association" to "causal association", helping a researcher understand what information we can gain from examining social-behavioral relations. Since a cross-match of relations among the domains is necessary for gaining information in research findings, then cross-match of the features of those relations becomes the foundation of the potential information gain. Brinberg and McGrath (1985) conclude that the relation within a given domain that has the fewest features to cross-match will constrain the character of the analysis, i.e. the weakest feature set will constrain analysis to the level of that feature set. That is why it is important to understand how the three research domains support and at the same time constrain each other. Examining the cross-matches for relation features will help sort out the flaws in research in a systematic way. Clearly, it is time consuming, and will take some effort. However, since the purpose of correspondence analysis is to explore potential information gain that can occur in a data analysis, then it seems like a useful undertaking in order to determine whether a research study will potentially find out anything new.

The reader can see that the analysis strategies in Table 4.8 address feature correspondence to varying levels (remember this is from the perspective of the methodological domain only). Clearly, some analysis strategies are more robust than others in terms of potential information gain. The higher the feature level attained according to the ten features described above, the more detailed the information gain. Thus, having information about the direction and strength of how map use is related to decision table use, is "more" information than just knowing that they are related to each other in some way. Remember that the ranking of relation features describes the specificity of a single relation, not the number of relations treated. Consequently, a researcher cannot conclude from this discussion of correspondence anything about a total information gain. If in a given study, a large number of level 1 features are identified as associated with the respective large number of relations; and in another study, only one relation is at issue but attains all seven tests, then only context of the study as specified by the research objectives would be able to provide an indication about which study generated more information. Thus, knowing that functional relationships exist between participant expertise and the amount of decision table or map use, might in fact be more information than knowing that there is a correlation among a wide array of decision aids. The nature of the analyses are different, hence the results will be as well. As one compares the listing of treatment modes with the level of information gain, there appears to be a trend indicating that higher numbers of treatment modes are associated with more information gain potential. Again, we emphasize that information gain is by relation, and not total gain for a particular study. Nonetheless, this examination of analysis strategies by potential information gain is a new contribution to understanding the difference in analysis strategies as one phase among many in research. Once an analysis strategy is planned and then carried out, a researcher can pursue the research strategy phase called "reporting strategy" to share research findings with an audience.

### 4.3.5 Reporting strategy

The reporting strategy phase is as important as any of the other phases in research. Scheduling, presentation style (audience), and robust comparisons of findings are among some of the most significant issues. We mention the first two and discuss the third in more detail, given the existence of treatments elsewhere. One significant aspect of reporting is the scheduling of the distribution of material. Clearly some topics are more time-sensitive than others. Substantive domain-lead (applied) research is perhaps the most time-sensitive of all, because such research is tied to the current contextual situation of people within their organizations interacting in a current political, social and economic climate. As conceptual domain-lead (basic) research has a longer "shelf-life" because it focuses on clarification concepts, thus early report distribution is not quite as critical. Methodological domain-lead research fits between the timeframes of the others. Reporting on methods is a contribution to "research infrastructure", but timely reporting of new methods can make a big contribution in a fast growing subject area — a characteristic of PGIS research. Because our PGIS research has involved applied, basic, and method-driven research orientations we have tried to balance all of those issues.

A second significant aspect of reporting is the presentation style of the results. The outlets for PGIS-related research are numerous. Style of reporting is linked to the orientation of research as guided by the needs of the audiences being considered. In our own work we have tried to balance among the three (applied, basic and methodological) orientations, and hence have used various outlets for work. As described earlier in this chapter, applied, basic, and methodological research comes about from researchers leading interests in the substantive, conceptual, and methodological domains. However, it is also important to remember that all three domains are part of every research study no matter what the orientation. The difference comes about as a matter of emphasis of domain, as well as how explicit/implicit influences are recognized from each domain. This book was written as an attempt to report on work in all three domains in somewhat equal balance because that is the perspectives we carry as researchers. We recognize how difficult it is to strike such a balance. No matter how much our presentation might span all three domains, we tend to emphasize the conceptual and methodological domains. Even with our practical interests in the substantive domain (as demonstrated by our several years' commitment to bring practical software to PGIS through http://www.geochoice.com), we recognize that this book takes more of a basic and methodological research orientation than it does an applied orientation. Chapters 2, 3, and 4 speak to the conceptual and methodological domains directly; whereas Chapters 5, 6 and 7 address a range of substantive issues using applied and basic orientations.

A third major aspect of reporting, and one that particularly interests us at this time, involves comparisons of findings in research reports. It is a concern that has little visibility to date with PGIS research, leaving the research open to criticism about systematic knowledge building, relying more on anecdotal knowledge building. In any single research study, the information gained through analysis has developed as challenges (successes and failures) in the various phases of research. The challenges (as threats to validity of research) occur across all phases no matter what the orientation of a study — no research is immune. Documenting the challenges to provide an even balance of reporting is useful to others who can learn from well-documented research experiences. How those findings are documented depends of course on the needs of the particular study. However, there is a generalization to be made here. Such findings and the way they are reported can, and should, be useful to other researchers examining similar topics. In fact, full disclosure of findings in research would seem to bolster the quality and hence publishability of research findings. Such disclosure enhances the potential of links to other findings. However, the links do not just start in the reporting phase. Yin (1994) makes a special point about documenting "data collection protocols" in case study research. A protocol for data collection, and the database that results, is a different contribution than the resultant findings synthesized from that database, but nonetheless almost as important. He suggests that a well-documented protocol will allow others to track the collection of data in a similar manner as is often reported in experiments. Thus, case study research can be viewed as systematic, and replicable. The resulting database can be re-analyzed in a similar or different manner in an attempt to corroborate, extend, or reinterpret findings by other researchers. The test of quality called "reliability" focuses on this issue of replicability of results. A well-documented protocol makes the overall research strategy more replicable.

Along with a reporting of findings from analysis, Brinberg and McGrath (1985) suggest that findings be compared to other findings as the core of stage 3 research, remembering that stages 1 and 2 are "planning" and "doing the research", respectively (see Table 4.3). This comparison with other findings in stage 3 actually has both an interesting parallel to meta-analysis research and an interesting difference. In meta-analysis research a researcher examines a set of findings on a particular topic in-depth, usually constrained to similar research design, and then attempts to synthesize those findings. The search for those findings is similar to a search in a good literature review performed for any study. However, a literature review for a proposed study does not commonly limit itself to a particular research design. The reason being is that a researcher would want to understand what has been found out about a topic from various research designs. The materials in the literature review thus contextualize the need for new information in the proposed study. If a researcher wanted to introduce "validity as robustness" early on in a proposal, then they would undertake such meta-analysis from the review findings. However, with a proposed study, and perhaps numerous research designs and findings being identified, it is rather time consuming to perform a meta-analysis for each of the designs. Consequently, syntheses from literature reviews commonly stop short of such in-depth analyses because of time constraints. Nonetheless, after a proposed study is undertaken, a researcher should return to the related findings identified in the literature review to compare and contrast the current findings with those appearing in the literature. As mentioned previously, stage 3 is where the most "knowledge building" can occur. That is, information gained from current findings are compared and contrasted against information gained from other findings. The two (or more) sets of information are synthesized to develop a better knowledge of the PGIS topic under consideration.

Now that we have evaluated each of the research strategy phases, we turn to a more specific investigation about what can be done to avoid (or at least reduce) threats to the quality of research findings. We do this through what Yin (1994) has called tactics relevant to different research strategy phases.

## 4.4 Tactics for enhancing the potential quality of research findings

When a research strategy is be used in a particular study, a researcher would need to address the detailed character of each of the phases of the research (design) strategy, evaluating the choices among treatment modes, data gathering strategies, and analysis strategies in regards to that study. The list of research strategies provided in Table 4.4, and the components for each phase in Tables 4.5–4.8 are at least a contribution toward that evaluation. The form of questions on the left side of Table 4.4 provide an initial filter for selection. As mentioned previously, our research concern has been with questions of "how and why" because we are interested in testing theory. However, a reader will find other questions addressed in Chapters 5–7 as well. When considering the choice of a component for a phase, we are dealing with how the subsequent phases to research questions influence, hence threaten, the potential quality of the findings for those questions.

The non-correspondence of the features for relations across the domains lead to threats to the quality of the design, hence overall strategy, and thus potential findings as a consequence of using a particular research strategy. As mentioned previously, various authors have addressed the nature of threats for particular research strategies, e.g. Cook and Campbell (1979) for experiments and quasi-experiments, Brinberg and McGrath (1985) for experimental design, and Yin (1994) for case studies. However, only Yin (1994) has provided a set of *tactics* for addressing the threats in terms of tests for potential quality of research design. By combining the tests for research quality together with the correspondences we can come to a better understanding of the threats to quality that exist. It would be extremely useful to do this for every research strategy listed in Table 4.4. However, space and time limits allow us only to focus on the strategies relevant to our past and current work (see Tables 4.9–4.11 ).

Our contribution here is to lay out the threats and a sample of respective tactics according to phase of research strategy. Furthermore, we recognize that both the conventional approach to quality tests (listed in column 2 of Tables 4.9–4.11) and the correspondence approach to quality (listed in column 3) are both useful for describing the threats. The tactic for addressing the threats are listed in column 4. The difference in presentation here makes understanding the challenges in quality somewhat straightforward — at least more so than in Brinberg and McGrath's (1985) and Yin's (1994) work.

Based on the evaluation and choice of research strategies above, and as a prelude to what is to come in subsequent chapters, it is easy for us to share with the reader an overview of how we have conducted the studies reported in Chapters 5, 6, and 7 of this book (see Table 4.12). The studies are ordered in increasing level of sophistication in their ability to address research questions, i.e. what, how and why. In a general sense the amount of time for set up and implementation parallels that increase in sophistication. As the reader will see, the strategies are different, hence the findings are different. Before we turn to those studies to share our findings, let us recap what we have found in this chapter.

## 4.5  Conclusion

This chapter could have been rather shorter if we relied solely upon the texts written about research methods. After all, there are so many to choose from, we should just be able to cite them and be done with all of this "research method stuff". Given one, two or three research strategies, the task of making use of them throughout a research career becomes a rather easy undertaking. But what if, in a collaborative situation, one researcher's three strategies are not the same as the second researcher's three strategies? This chapter was written to seek clarification of the complex state of affairs in research methods from a broad-based perspective. We treated 18 research strategies, a sufficiently large number to provide something for just about everyone. In doing so, we want to open the door for new challenges related to social-behavioral studies about PGIS use.

*Table 4.9* Tactics for establishing potential quality in research findings for a laboratory experiment

| Phase of research | Criteria for test of quality | Research (validity) stage (Brinberg and McGrath 1985) | Laboratory experiment quality tactics |
|---|---|---|---|
| research design | external validity | – stage 1 strategic considerations for research setting, choose actors, place and time as appropriate | – choose a sample of the population to which you want results to apply |
| data collection | construct validity | – stage 2 correspondence between concept and method to make concept operational<br>– stage 1 identify the elements and relations that are significant | – ask multiple subjects, independently about data coding interpretation<br>– use focus group for interpretation of data coding results |
| data collection | reliability | – stage 3 replication robustness – know how data were created | – when using multiple coders, use Kappa reliability check on coding |
| data analysis | internal validity | – stage 2 correspondence cross-domain match of method and substance for relation at the level of feature 1, i.e. a locational or attribute relation exists between $i$ and $j$; and perhaps avoiding the influence of features 8–10 of external elements | – show that the independent and dependent variables follow from the theory |
| data analysis | internal validity | – stage 2 correspondence validity at the level of feature 3, i.e. $i$ leads to $j$ as causal link;<br>– stage 2 features 6–10 demonstrate internal and external stability | – perform multiple method analyses to uncover dynamics |
| composition | construct validity | – stage 3 validity as robustness verification using participants' perspective | – what have other experiments found? |

*Table 4.10* Tactics for establishing potential quality in research findings for a field experiment (quasi-experiment)

| Phase of research | Criteria for test of quality | Research (validity) stage (Brinberg and McGrath 1985) | Field-experiment quality tactics |
|---|---|---|---|
| research design | external validity | – stage 1 strategic considerations for research setting, choose actors, place and time as appropriate | – choose a sample of the population to which you want results to apply |
| data collection | construct validity | – stage 2 correspondence as description of the features in each domain | – make sure theoretical explication is sufficient<br>– use focus group to validate data representativeness |
| data collection | construct validity of cause and effect are different | – stage 2 correspondence between concept and method to make concept operational<br>– stage 1 identify the elements and relations that are significant | – use multiple methods for measuring cause and effect variables |
| data collection | statistical conclusion validity – covariation in the variables | – stage 3 replication robustness – know how data were created | – large numbers of samples<br>– replicable treatments<br>– measurement before treatment, measure when treatment applied, measure after treatment |
| data collection | statistical conclusion validity | | – make sure measurement is taken before treatment, and when treatment is applied |
| data analysis | internal validity | – stage 2 correspondence cross-domain match of method and substance for relation at the level of feature 1, i.e. a locational or attribute relation exists between $i$ and $j$; and perhaps avoiding the influence of features 8–10 of external elements | – show that treatment preceded measured effect |
| data analysis | internal validity | – stage 2 correspondence validity at the level of feature 3, i.e. $i$ leads to $j$ as causal link;<br>– stage 2 features 6–10 demonstrate internal and external stability | – make sure subjects are assigned randomly to treatments |
| composition | construct validity | – stage 3 validity as robustness verification using participant's perspective | – what have other experiments found? |

*Table 4.11* Tactics for establishing potential quality in research findings in a case study

| Phase of research | Criteria for test of quality | Research (validity) stage (Brinberg and McGrath 1985) | Case study quality tactics (after Yin 1994) |
|---|---|---|---|
| research design | external validity | – stage 1 strategic considerations for research setting, choose actors, place and time as appropriate | – use replication logic in multiple case studies |
| data collection | construct validity | – stage 2 correspondence between concept and method to make concept operational<br>– stage 1 identify the elements and relations that are significant | – use multiple sources of evidence as in several data collection techniques to measure a variable in multiple ways |
| data collection | construct validity | – stage 2 correspondence as description of the features in each domain | – establish chain of evidence explaining what led from what |
| data collection | reliability | – stage 3 replication robustness – know how data were created | – use case study protocol |
| data collection | reliability | – stage 3 replication robustness – know that data exist in some systematic form | – develop case study database |
| data analysis | internal validity | – stage 2 correspondence cross-domain match of method and substance for relation at the level of feature 1, i.e. a locational or attribute relation exists between $i$ and $j$; and perhaps avoiding the influence of features 8–10 of external elements | – do pattern matching between multiple variables, not necessarily just independent and dependent |
| data analysis | internal validity | – stage 2 correspondence validity at the level of feature 3, i.e. $i$ leads to $j$ as a causal link;<br>– stage 2 features 6–10 demonstrate internal and external stability | – do explanation building |
| data analysis | internal validity | – stage 2 correspondence validity at the level of feature 2, i.e. $i$ precedes $j$ as a temporal relation | – do time-series analysis |
| composition | construct validity | – stage 3 validity as robustness verification *using participant's perspective* | – have key informants review draft case study report |

*Table 4.12* Research strategies for studies reported in Chapters 5, 6, and 7

| Chapter: strategy name | research question | research design data strategy* | analysis strategy | reporting strategy |
|---|---|---|---|---|
| Chapter 5: Application description task analysis | – what aspects | – task description coding | – task analysis | – emphasize aspects in each construct |
| Chapter 6: Case study proposition analysis | – why relationship between aspects | – case document coding | – case proposition analysis | – emphasize relationships among aspects |
| Chapter 7 HCHI laboratory process analyisis | – what, when relationship between aspects | – laboratory experiment process coding | – nonparametric inferential ESDA – parametric inferential ANOVA and regression | – emphasize process of changes in codes for aspect – emphasize occurrence among aspects |

\* Includes treatment mode as first part of this strategy.

Much of this chapter was motivated by the work of Joseph McGrath and his co-authors, who have been concerned about validities of empirical research findings and systematic and comprehensive frameworks to address them (Brinberg and McGrath 1985, Runkel and McGrath 1972, McGrath 1995). This chapter provided us with an opportunity to re-examine what we have learned earlier, and come to grips with what different approaches to social-behavioral research have us understand about research findings. In the same spirit of Eisenhart and Borko (1993) we seek to elucidate guidelines for recognizing "good" empirical research. Research that meets multiple validities, but agreed upon validities. We seek to improve the current state of methods in PGIS studies, so often characterized as anecdotal research, although anecdotal research is sometimes the bud of a blossoming research agenda. Such anecdotes carry us only so far toward knowledge development, i.e. knowledge about PGIS in society.

A growing trend in public–private participatory activity suggests that complex public–private problems, e.g. growth management-oriented land use planning and development, are likely to be addressed through inter-organizational strategies at local, regional, and federal levels of decision making. PGIS technology is likely to get used in a more strategic way in support of various parts of the public–private decision making process; this is because more and more PGIS information is proving to be useful to a wide variety of groups across society, including technical· specialists, decision makers, and interested and affected parties. How PGIS techno-logy will develop to support analysis and deliberations within group set-tings can be better understood through social-behavioral studies about PGIS use.

The research reported in this chapter was motivated by a challenge to connect research questions to treatment modes, treatment modes to data gathering strategies, data gathering strategies to data analysis strategies, and data analysis strategies to reporting strategies, in a more effective way. That connection contributes to a broad research agenda about PGIS use and the social implications such use has for society. To make that connec-tion more systematic, we have sorted through the apparent chaos in research methods terminology, particularly that involving social-behavioral research methods. Even some of the most complete treatments of social-behavioral research methods in literature (McGrath 1982, Brin-berg and McGrath 1985, McGrath 1995) leave room for a better articula-tion. Starting out with a clearer sense of the connections between research questions and data gathering strategies in stage 1 (planning the research) is better than waiting until stage 2 (doing the research) to recognize the correspondence among elements and relations, particularly when address-ing problems that are complex and multidisciplinary. Knowing what the implications of stages 1 and 2 are for the knowledge building that can occur in stage 3 (corroborating the research) provides for a well-informed research process.

Research strategy is an overarching term that includes problem articula-tion, treatment modes, data gathering strategy, data analysis strategy, and reporting strategy. Our contribution concerns how we can choose among research strategies to address a research question, while being fully aware that we are striking a balance among the substantive, conceptual, and methodological domains of research. We now also understand that when one of those domains is used to lead the research charge, we automatically establish the orientation of the research as applied, basic and method-driven, respectively. Furthermore, that orientation takes on an interesting character when we choose the second domain to support the first. By fiat, the third domain follows and in many instances constrains the nature of the findings. Thus, the lesson here is "watch for the third domain", embrace it and use it rather than allow it to use you.

It is clearer for us how the array of social-behavioral strategies provide a wealth of opportunities for pursuing empirical studies about use of GIS in general, and PGIS use in particular. Because all strategies have shortcom-ings, sorting through the choices of when to use what strategies is import-ant. In this chapter we developed a number of tables and figures that help us compare and contrast the strategies. We hope this view can help other researchers as well, even if PGIS technology use is not the chosen topic in the substantive domain as it is for us. However, we admit the new view is not revolutionary, it is evolutionary. We stand on the shoulders of those who have gone before. Since research is a knowledge accrual process we gladly recognize those who have elucidated the material we used in order for us to articulate this new view.

Our frameworks clearly point out that no single strategy can cover all the bases for studying PGIS use in particular, as well as advanced information technology use in general. Our chosen theoretical domain called EAST2, with the premises helping us to focus our research questions, indicates that a variety of strategies are appropriate for examining the myriad of concerns about PGIS use. By balancing a substantive perspective (domain) with a conceptual perspective (domain), together with a methodological perspective (domain), we believe the PGIS research community can make progress with theoretical and methodological turns in geographic information science as suggested by Pickles (1997).

The theoretical and methodological turns in GIS are needed to keep up with the substantive turn GIS made some time ago, by virtue of its growing popularity worldwide. In our research we are seeking to bolster the theoretical and methodological contributions related to our interests in group decision making about environmental cleanup and habitat restoration, transportation, land use, and public and environmental health. Research that does not treat all three — substantive, conceptual, and methodological — domains explicitly in any given study provides only partial insight about the nature of the research contribution. As it is important to know that any variable ignored implicitly in a study can confound the empirical results in the study, a domain that is ignored implicitly may also confound the knowledge accrual process in research. We recognize that a full-fledged contribution from all three domains is not likely to be made by any single research study. In fact, a full-fledged balance should be sought not for a single study, but across connected research studies. As such, knowledge accrual through systematic application of complementary research studies based on complementarity across domains, as well as within domains, is possible. Mixed-method research is the basis of this complementarity. What the research strategy frameworks provide are a basis for developing a better understanding of mixed-strategy (method) research. We advocate the use of complementary strategies to advance knowledge accrual about PGIS use through mixed-method research.

The research strategy frameworks developed here provide only general guidelines. The tables and figures in this chapter demonstrate clearly the advantages and disadvantages of research strategies as well as the complementarity of research strategies. We hope that the strategy frameworks reported in this chapter provide the basis of a robust framework to assist researchers with their choices among strategies, and that the latter, as presented here, are made clearer and therefore easier for others as they are for us. Articulating better research strategies, hence writing better research proposals, undertaking the research, and moving toward solutions for complex problems, depend on making the choices as clear as possible. However, only time, research reporting, and/or a much more thorough treatment of the material here will elucidate whether strategies alone or in combination are the better ones for researching PGIS use as contributions to participatory geographic information science. Elaborating issues about the choices for research strategy can be a book in itself, so here we are encouraged to provide at least an overall synthesis of some of the trade-offs researchers face when performing social-behavioral research about PGIS use. Perhaps we or others reading this material might be encouraged to write such a book. Although a single chapter cannot do justice to the entire topic of research strategies, the absence of such material should not stop us from pursuing interesting research studies about PGIS use. We report on such studies in the following chapters.

# References

Avison, D., Lau, F., Myers, M. and Nielsen, P. (1999) "Action research", *Communications of the ACM*, 42(1): 94–7.

Brinberg, D. and McGrath, J.E. (1985) *Validity and the Research Process*, Thousand Oaks, Sage.

Cook, D.T. and Campbell, D.T. (1979) *Quasi-Experimental Design: Design and Analysis Issues for Field Settings*, Skokie, IL, Rand McNally.

Densham, P.J., Armstrong, M.P. and Kemp, K. (1995) "Report from the Specialist Meeting on Collaborative Spatial Decision Making", Initiative 17, National Center for Geographic Information and Analysis, UC Santa Barbara, 17–21, September 1995.

Drew, C., Nyerges, T., McCarthy, K. and Moore, J. (2000) *Using Decision Chains to Explore Four DOE-EM Decisions: A Cross-Case Analysis*, CRESP Technical Report, Consortium for Risk Evaluation with Stakeholder Participation, Rutgers University and University of Washington.

Eisenhart, M. and Borko, H. (1993) *Designing Classroom Research*, Boston, Allyn and Bacon.

Fisher, P. (1998) "Geographic information science and systems: a case of the wrong metaphor", in *Proceedings*, 8th International Symposium on Spatial Data Handling, eds. Thomas K. Poiker and Nicholas R. Chrisman, Simon Fraser University, Department of Geography, International Geographical Union, Geographic Information Science Study Group.

Goodchild, M.F., Egenhofer, M.J., Kemp, K.K., Mark, D.M. and Sheppard, E. (1999) "Introduction to the Varenius project", *International Journal of Geographical Information Science*, 13(8): 731–45.

Jankowski, P. and Nyerges, T.L. (2001) "GIS-supported collaborative decision making: results of an experiment", *Annals of the Association of American Geographers*, in press, March issue.

Jankowski, P., Nyerges T.L., Smith, A., Moore, T.J. and Horvath, E. (1997) "Spatial group choice: a SDSS tool for collaborative spatial decision-making", *International Journal of Geographic Information Systems*, 11(6): 577–602.

Jankowski, P. and Stasik, M. (1997a) "Design consideration for space and time distributed collaborative spatial decision making", *Journal of Geographic Information and Decision Analysis*, 1(1): 1–8.

URL: http://publish.uwo.ca/~jmalczew/gida.htm.

Jankowski, P. and Stasik, M. (1997b) "Spatial understanding and decision support system: a prototype for public GIS", *Transactions in GIS*, 2(1): 73–84.

Kidder, L. and Judd, C.M. (1986) *Research Methods in Social Relations* (5th edn), New York, Holt, Rinehart and Winston.

Mark, D. (ed.) (2000) "Geographic information science: critical issues in an emerging cross-disciplinary research domain", *URISA Journal*, 12(1): 45–54.

McGrath, J.E. (1982) "Dilemmatics: the study of research choices and dilemmas", in J.E. McGrath, J. Martin and R.A. Kulka (eds) *Judgement Calls in Research*, Beverly Hills, Sage: 69–102.

McGrath, J.E. (1995) "Methodology matters: doing research in the behavioral and social sciences", in R. Baeker, J. Grudin, W. Buxton and S. Greenberg (eds) *Readings in Human-Computer Interaction: Toward the Year 2000*, San Francisco, Morgan-Kaufmann Publishers: 152–69.

Miller, R.P. (1995) "Beyond method, beyond ethics: integrating social theory into GIS and GIS into social theory", *Cartography and Geographic Information Systems*, 22(1): 98–103.

Nyerges, T., Moore, T.J., Montejano, R. and Compton, M. (1998a) "Interaction coding systems for studying the use of groupware", *Journal of Human-Computer Interaction*, 13(2): 127–65.

Nyerges, T., Montejano, R., Oshiro, C. and Dadswell, M. (1998b) "Group-based geographic information systems for transportation site selection", *Transportation Research C: Emerging Technologies*, 5(6): 349–69

Onsrud, H.J., Pinto, J.K. and Azad, B. (1992) "Case study research methods for geographic information systems", *URISA Journal*, 4(1): 32–44.

Orlikowski, W.J. (1992) "The duality of technology: rethinking the concept of technology in organizations", *Organization Science*, 3(3): 398–427.

Pasquero, J. (1991) "Supranational collaboration: the Canadian environmental experiment", *Journal of Applied Behavioral Science*, 27(1): 38–64.

Pickles, J. (ed.) (1995) *Ground Truth: The Social Implications of Geographic Information Systems*, New York, Guilford.

Pickles, J. (1997) "Tool or science? GIS, technoscience, and the theoretical turn", *Annals of the Association of American Geographers*, 87(2): 363–72.

Runkel, P.J. and McGrath, J.E. (1972) *Research on Human Behavior*, New York, Holt, Rinehart and Winston.

Sackmann, S.A. (1991) "Uncovering culture in organizations", *Journal of Applied Behavioral Science*, 27(3): 295–317.

Sanderson, P.M. and Fisher, C. (1994) "Exploratory sequential data analysis: foundations", *Human-Computer Interaction*, 9: 251–317.

Scarbrough, E. and Tanenbaum, E. (eds) (1998) *Research Strategies in the Social Sciences*, Oxford, Oxford University Press.

Sheppard, E. (1995) "GIS and society: towards a research agenda", *Cartography and Geographic Information Systems*, 22(1): 5–16.

Sheppard, E., Couclelis, H., Graham, S., Harrington, J.W. and Onsrud, H. (1999) "Geographies of the information society", *International Journal of Geographical Information Science*, 13(8): 797–823.

Stasik, M.I. (1999) *Collaborative Planning and Decision Making under Distributed Space and Time Conditions*, unpublished doctoral dissertation, University of Idaho.

Williams, F., Rice, R. and Rogers, E. (1988) *Research Methods and the New Media*, New York, The Free Press.

Wright, D., Goodchild, M. and Proctor, J. (1997a) "GIS tools or science?: demystifying the persistent ambiguity of GIS as 'Tool' versus 'Science'", *Annals of the Association of American Geographers*, 87(2): 346–62.

Wright, D., Goodchild, M. and Proctor, J. (1997b) "Reply: Still hoping to turn that theoretical corner", *Annals of the Association of American Geographers*, 87(2): 373.

Yin, R.K. (1993) *Applications of Case Study Research*, Thousand Oaks, Sage.

Yin, R.K. (1994) *Case Study Research*, 2nd edn, Thousand Oaks, Sage.

Zigurs, I. (1993) "Methodological and measurement issues in group support systems research", in L.M. Jessup and J.S. Valacich (eds) *Group Support Systems*, New York, Macmillan Publishing.

Zmud, R.W., Olson, M.H. and Hauser, R. (1989) "Field experimentation in MIS research", *Harvard Business School Research Colloquium*, Publishing Division, Harvard Business School, Boston, MA: 97–111.

# 5    Collaborative spatial decision making in primary health care management

## A task analysis-driven approach

### Abstract

An important problem in addressing the primary health care needs of underserved rural areas is the allocation of financial resources. In this chapter we report on the use of a spatial decision support system used by a group of health care decision makers within the Department of Health and Welfare of the State of Idaho. Distributing limited financial resources in an equitable, yet need-responsive way is an issue faced by health management agencies not only in Idaho, but also across the USA. The funding allocation problem has a significant spatial component. The decision on which counties should receive funds is driven by the location of counties. Location determines to a large degree the distribution of available health care resources and consequently the coverage of health care needs. The decision problem can be characterized as multicriterion and evaluative, since its closure requires most certainly the evaluation of health care needs in Idaho counties based on a number of attributes. Which of these attributes should be selected as effective evaluation criteria is a question that becomes part of the decision problem. A decision support tool called GeoChoicePerspectives was used to assist in the allocation of funds. A task analysis method was used to describe the system requirements needed by decision makers to carry out the geographic decision support. Capabilities that portray maps and decision tables are described. Other advanced decision support techniques are used to describe the potential for group decision support.

The delivery of health care services in rural areas of the USA has been traditionally a challenge for state agencies responsible for primary health care. Low population density and high spatial dispersion of human settlements, coupled with the physician-to-population ratios well below the US national average, contributed to persistent shortages of primary health care services in many rural areas of the country. The State of Idaho is one example of this problem. As a predominantly rural state with a population density of 12.9 persons per square mile, and only 43.8% of the state population living in towns larger than 5,000 people, Idaho has struggled to cover adequately the need for primary heath care services of its rural residents. A preliminary study conducted in 1993 by the Center for Vital Statistics of the Idaho Department of Health and Welfare revealed that 8.2% of the demand for primary health care services in Idaho was unmet, due to spatial distribution of primary health care providers favoring urbanized areas and neglecting rural areas of the state. In 1994 the state had three counties with no resident physician at all (Center for Vital Statistics 1994).

At the state level, an important problem in addressing the primary health care needs of underserved areas is the allocation of financial resources. How to distribute limited financial resources in an equitable, yet need-responsive way is an issue faced by health management agencies not only in Idaho but also across the USA. The problem has a significant spatial component. The decision of which counties should receive funds is driven by the location of counties. Location determines to a large degree the distribution of available health care resources and consequently the coverage of health care needs. The decision problem can be characterized as multicriterion and evaluative since its closure requires most certainly the evaluation of health care needs in Idaho counties based on a number of attributes. Which of these attributes should be selected as effective evaluation criteria is a question that becomes part of the decision problem.

Depending on the extent of knowledge on how to proceed with the evaluation process, a spatial decision problem, such as this one concerning allocation of funds for primary health care, can be structured or partially structured. An example of a structured problem involves a situation where all evaluation criteria are established and data describing the performance of areal units on these criteria are available. The problem is partially structured if evaluation criteria are incomplete and/or not all data are readily available. Regardless of whether the problem is structured or only partially structured, its multi-attribute character demands information processing capabilities challenging the cognitive abilities of most decision makers. Consequently, its decision activity may benefit from employing a spatial decision support system. Furthermore, not only is this problem rather complex for an individual decision maker, it is further complicated because it has a significant collaborative component. Typically, the decision process involves the collaboration of three very different perspectives: a group of health district representatives and health care providers (stakeholders), health agency officials put in charge of managing health care budget (decision makers), and health policy analysts fulfilling the role of technical experts. The allocation of funds is based on demonstrated need for services, using either a previously agreed formula and/or as a result of a negotiation process. Such a process may involve some form of cooperation, communication, and coordination of efforts by stakeholders and decision makers, in which case it becomes *de facto* a collaborative decision making activity.

In this chapter we discuss how a task analysis of a decision situation can be used to structure/prepare a collaborative approach to deal with a health care funding allocation problem. In order to set the context for a group decision approach, section 5.1 presents the background for task analysis concerning the statewide allocation of funds for primary health care services. Following this, a scenario involving a collaborative approach to allocating funds for primary health care services is presented in four subsequent sections; section 5.2 focuses on decision option generation, while sections 5.3 and 5.4 present the issues of criteria identification and option evaluation, respectively. Section three presents a brief discussion of the option generation task. Sections 5.3–5.5 provide a detailed task analysis of the current approach and contrast it with a future potential approach to group decision making concerning decision criteria identification and option evaluation. The chapter ends with our conclusions about a task-analysis based approach to group decision support for developing funding decisions concerning primary health care services.

## 5.1 Health care funding allocation decisions: a task analysis

The decision strategy of allocating funds for primary health care services to Idaho counties, is comprised of three phases: (1) *decision option generation,* (2) *criteria identification*, and (3) *decision option evaluation*. Before a collaborative decision support system can be created, a thorough analysis of constructs and aspects of the decision situation is needed (see Chapter 2). Identifying and documenting various aspects of a decision situation, particularly the aspects that influence the decision strategy, and its constituent decision phases, is called task analysis. The objective of task analysis is to address systematically decision support needs before they arise in practice. So in essence, task analysis fulfills the role of a planning process for collaborative spatial decision making. In the task analysis each decision phase can be treated as a separate task or, alternatively, they can all be treated as one decision task. The latter approach is justifiable only if all aspects of the task analysis are the same across the phases. In the decision problem at hand, some of the constructs for the option generation phase differ from the constructs of criteria identification phase and option evaluation phase. Hence, in the remainder of the chapter we will analyze the decision support needs of each phase separately by addressing their convening, process, and outcome constructs (see Chapter 2). Note that the three macro-phases used for describing the macro-micro approach in Table 3.1 are the same, but the ordering of the macro-phases in this decision situation is different because the situation is different. This change in ordering supports our contention that macro-phases are but phases of an "agenda" as established by participants addressing a particular situation.

Any collaborative decision situation (and each task within it) can be analyzed in respect to its convening constructs, decision process constructs, and decision outcomes constructs. The *convening* constructs articulate what is important in setting up a decision task, e.g. the people from organizations that are to participate, values, goals, and objectives of those participants and their organizations, and the information technology that can be made available to support the decision process. The *process* constructs include the dynamics of invoking decision aids, managing decision tasks from phase to phase, and the emergence of information structures such as maps, models, and databases. The *outcome* constructs and associated aspects include direct outcomes related to the specific decision task, and the social relations created, evolved, and/or destroyed when the task is completed. Before we present the analysis of decision tasks by constructs, we provide an overview of the decision problem.

Idaho is a predominately rural state with numerous geographic and access barriers. Sixteen of the state's 44 counties are considered frontier (less than six persons per square mile). Using the definition of urban to be a county with at least one population center having more than 20,000 persons, there are only seven urban counties in Idaho. Twenty counties have no population center of 20,000 or more, but average six or more persons per square mile. Sparsely populated areas, vast desert and numerous mountain ranges complicate the delivery of primary care services throughout the state.

Between 1980 and 1990, the elderly population in Idaho grew more rapidly than other age groups in the state, and faster than the national average growth in this population group. A total of 12.1% of Idaho's population was aged 65 years old or older in 1990, compared to 9.9% in 1980. The aging of the state's population will place an increasing demand on primary care services over the next decade.

Idaho has historically claimed one of the lowest physician-to-population ratios in the USA. Currently, Idaho ranks 48th in the USA, with 125 physicians in patient care per 100,000 population (Jankowski and Ewart 1996). Recruiting health professionals to Idaho's remote areas and to regions in the state with substantial portions of low-income/underserved persons is a constant challenge. The Idaho Department of Health and Welfare (IDHW) completed a survey of Idaho's rural community facilities (excluding Boise, Nampa, Caldwell, Idaho Falls, Lewiston and Coeur d'Alene) in October 1995. The results of the survey indicated that 28 of the 47 facilities interviewed were actively recruiting primary care physicians or midlevel providers. The top two reasons for recruitment were increased demand and to relieve the workload of other primary care providers. The analysis completed by IDHW in 1994 indicated a maldistribution of primary care providers, with an adequate concentration of physicians and midlevel providers in the more populated areas of the state and inadequate numbers in the more rural and frontier areas. The state Health Professional Loan Repayment Program and the Health Professional

Clearinghouse have been working to alleviate the challenges that many rural/underserved Idaho communities face when recruiting health care providers. The decision problem they face is how to distribute limited funds (US$250,000) in order to help the state counties attract health care professionals through repaying their education loans. The distribution of funds among the state counties should be equitable, i.e. based on demonstrated needs, and provide a high likelihood of successful hiring and longer-term retention of health care professionals. It is a shared need for action relevant to the overall decision problem that brings the participants together in this collaborative undertaking.

## 5.2 Decision option generation

A curious reader may ask about the meaning of a decision option in the problem discussed here. In a non-spatial situation the decision option represents a feasible decision variant, e.g. a specific technological variant from a range of feasible technological solutions or a product from a number of feasible products. In a geographic decision situation the decision option is typically identified with a specific location. Depending on the spatial scale of a problem, a location representing the decision option can be represented by point (e.g. site), area (e.g. county), line (e.g. water pipeline corridor) or any combination of the above such as in the case of a land use plan. Consequently, the decision option generation involves finding locations that meet suitability criteria. There are two approaches to the task of finding suitable locations: *choice* and *design*. Choice involves the selection of suitable locations from the set of all possible discrete locations. In the context of GIS the choice of suitable locations is usually accomplished by conjunctive or disjunctive attribute queries. Design involves the specification of suitable locations from a continuous space of locational possibilities. In the GIS context this is usually accomplished via one or both of Boolean overlay and/or Weighted Linear Combination (see section 3.3).

The *option generation* phase used frequently in other decision making strategies (e.g. macro strategy proposed by Renn *et al.* (1993) for solving environmental decision problems) is rather simple in this case. However, what makes this phase simple is the "institutionalization" of counties as a primary mechanism through which distribution of funds occurs. Because of the institutional and administrative constraints requiring the use of counties as the primary spatial units of analysis, the participants of the decision process adopted counties as decision options. Every county has to be evaluated and ranked, thus establishing a rank order on the set of counties. The rank order represents a priority list of counties in need of funding, from the highest to the lowest funding priority.

Because county governance and administration are highly institutionalized processes in local government in the USA, we expect the same institutional and administrative constraints to be the basis for generation of decision options in the future.

## 5.3  Criteria identification

A task analysis of the collaborative decision situation involving convening, process, and outcome constructs can be carried out by decision analyst(s) using the information obtained from the participants of a collaborative decision making process. The information can be obtained through discussions and interviews with participants conducted by the analysts. In the decision situation described here, the authors fulfilled the role of decision analysts.

### *5.3.1  Convening constructs of criteria identification*

The decision science literature advocates a hierarchical structure approach to organizing convening constructs of a decision situation, such as values, goals, objectives, and criteria (Saaty 1990, von Winterfeldt 1987, Keeny 1992). In this approach values are at the top of the hierarchy (or root of a tree, to adopt an alternative metaphor that means the same thing). Values represent something a person or a group deeply cares about. Organizations use mission statements and statutes to espouse their underlying values. Goals stem from values as concerns articulating a particular value in a given context. Objectives specify directions of attainment that are sought for a given goal. Criteria, resulting from objectives, are measurable characteristics expressing the degree to which these objectives are achieved for a particular option.

#### *5.3.1.1  Values, goals, objectives and criteria*

In the case of the Idaho health care funding decision situation, the value with which all the participants (hence organizations they represent) identify is:

* Health care is an important service to be provided to citizens of Idaho in order to maintain a healthy population.

The outcome goal of criteria identification phase, derived from this value and formulated by the participants of the decision making process is:

* Selection of criteria for the evaluation of health care needs in Idaho counties.

Since the need for health services is not a simple, one-attribute based measure, but rather a function of many attributes (criteria), the quantification of the need is a multicriterion evaluation problem, which depends on the selection of evaluation criteria. The outcome goal encourages the consideration of two objectives:

- Estimating the unsatisfied demand for health care services;
- Calculating the supply and spatial distribution of health care services for each county.

A criterion that can be used to measure the attainment of the first objective for every county in Idaho is the *estimated number of unaccounted primary care visits*. This criterion serves as an indicator of the unmet demand for primary care medical services due to the unavailability of providers within a given travel distance. The primary care providers include physicians with family practice, general practice, internal medicine, pediatrics, obstetrics/gynecology specialties, primary care nurse practitioners and physician assistants. The number of unallocated visits is an estimated count based on the number and geographic distribution of primary care providers, number of hours available by each primary care provider, population distribution in Idaho by census block groups, estimated primary care visits based on the age and sex of population, and travel distances between the primary care providers and the population locations.

The second objective can be initially measured by two criteria: *availability of on-call providers* and *proximity of a hospital to the population*. The first criterion is expressed by the number of hours on call for each provider. The number of hours on call is multiplied by the provider's full time employment (FTE) at each location. The results are summed for each county and the total is divided by the overall FTEs in each county. This yields the measure of the average call burden per FTE in each county. The second criterion is calculated based on the nationally accepted standard for the maximum travel distance of 35 miles to a hospital in rural areas. The number of individuals residing outside each hospital influence zone (35-mile buffer) is then calculated for each county.

So far, we have concentrated in the task analysis on goals, objectives and criteria concerning the decision situation "outcomes". This addresses the question of what is expected from the criteria identification phase of the decision process. A similar task structure comprised of goals and objectives can be analyzed with respect to the "process" of identifying criteria, which addresses the question of how outcomes are to be achieved.

There are two fundamental goals of the process of criteria identification identified by the participants:

- Ensure the equality of participation in the process; and
- Contribute to the credibility of the agency in charge of funding primary health care services (Idaho Department of Health and Welfare).

The objectives resulting from the first goal include:

- Equalize the standing of participants in the process;
- Give every participant an opportunity to express his/her concerns and viewpoints;
- Provide for personal choice of criteria.
- Provide for both open and anonymous participation in the process.

The objectives resulting from the second goal include:

- Ensure the equal representation of all counties in Idaho;
- Strive for consensus among the process participants regarding the adopted evaluation criteria.

No specific criteria measuring the achievement level of process objectives were identified. The objectives listed above act as "means" objectives contributing to better achievement of process goals, which in turn contribute to better achievement of the outcome goal. Commonly, process criteria are identified and measured in rather contentious decision situations, when a group is unsure about being able to articulate the potential options and/or outcomes, as in a number of environmental problems. Thus, the process criteria become the indicators of success. Such is not the case with rural primary health care funding because distribution of "dollars" makes sense to all participants involved, even if some of the exogenous, contextual factors are rather complicated to influence.

### 5.3.1.2 Participants involved

The participants involved in the criteria identification phase include stakeholders, decision makers, and technical experts. The stakeholders and decision makers are the members of the Board of Directors of Health Professional Loan Repayment Program. The Board is comprised of community leaders, hospital administrators, doctors and mid-level providers in Idaho. The technical experts include health policy analysts from Idaho Department of Health and Welfare's Center for Vital Statistics and Health Policy and spatial decision analysts from the University of Idaho.

### 5.3.1.3 Decision support tools

The choice of decision support tools depends on the meeting arrangement for collaborative decision making, budget and time limits, the experience of participants in the use of computer technology, their educational background, and the type of information to be considered in a given decision situation (see Chapter 3).

Two principally different meeting arrangements — conventional face-to-face meetings (same place/same time) and time-space distributed meetings (different place/different time) — can be considered for the criteria identification decision phase. Along these diametrically opposing meeting arrangements for collaborative decision making there exists a possibility of a mixed solution comprised of space-time distributed discussion, followed up by a conference-call meeting to finalize the selection of criteria. While both of the principal arrangement alternatives (same place/same time and different place/different time) require the presence of human facilitator/mediator, different software and computer network technologies are required in each case.

A meeting organized in the same place/same time environment can be performed in a conference room setting with a single computer connected to a large display and/or participant computers connected by a local area network. In the case of different place/different time arrangements, participants collaborate in a virtual space via their PCs connected by a wide area network, i.e. the internet. In either case, the software capabilities may include:

- anonymous input of ideas concerning the evaluation criteria;
- polling and display of textual ideas;
- support for argumentation and discussion of candidate criteria;
- intelligent software agents allowing the computation of impacts of a given criterion on competing decision options;
- electronic voting, compilation of voting results, facilities to identify the consensus-based criteria.

All of the participants had college education. However, their familiarity with computer technology varied from a sporadic use of personal computers (a few hours per month) to a daily routine (a few hours per day). Hence, decision support tools should be capable of accommodating different levels of human-computer-human interaction (from novice to expert user). Additionally, due to different levels of familiarity with computer technology, the group should consider retaining the services of group facilitator and chauffeur, in the case of meeting room environment, and a facilitator/mediator in the case of distributed space and time collaboration environment. Results of experimental research strongly suggest that the presence of group facilitator can enhance users' satisfaction with computer-supported collaborative decision process (Chun and Park 1998). The role of group facilitator is to provide guidance and mediation in the collaborative decision process, whereas the role of chauffeur is to provide technical support with the use of software. If the group is small the same person can fulfill the role of facilitator and chauffeur. As research on group decision support systems demonstrates, a skillful group facilitator can compensate for the lack of user experience with GDSS tools (Chun and Park 1998).

### 5.3.2 Process constructs of criteria identification

The process constructs of criteria identification phase help describe tool interaction during the collaboration. The constructs include:

- appropriation of tools: how tools are invoked to support the collaboration; and
- task management: how the criteria identification process is organized.

In contrast with a more traditional approach to criteria identification involving multiple decision makers (Keeney *et al.* 1994), where decision analysts interview each decision maker separately in order to elicit his/her value-goal preference structure, in the collaborative approach the participants work together, aided by a facilitator/chauffeur and spatial decision support software. The software/facilitator capabilities that are useful in supporting criteria identification include:

- Issue Development — combine inputs from various participants to generate the list of issues, goals, objectives, and criteria relevant to the decision problem;
- Issue Exploration — depict geographic outcomes of evaluation criteria and help participants develop a shared understanding of the outcomes;
- Issue Analysis — highlight differences, look deeper into criteria and be able to estimate the relationships between criterion values and the decision variable (i.e. the amount of funding allocated to each county).

The software tools implementing these capabilities can be invoked in two modes: a private/public mode or a public mode. In the private/public mode each participant has access to a networked computer with decision support software and problem-specific database. The participants have time allocated during the meeting for the individual exploration and analysis of issues surrounding the criteria development. The results of individual explorations are brought into the open by a facilitator during the public brainstorming and discussion stage of the meeting. In the public mode a facilitator manages the exploration, analysis, brainstorming, and coalescing of participants' ideas; the results are displayed on a single large screen that can be viewed by all participants. Regardless of whether the mode of collaboration is private/public or public only, the process is managed by the facilitator, who directs the participants to work on specific subtasks (i.e. development of goals, objectives, and criteria), gathers participants' input by means of voting, and organizes the discussion.

### 5.3.3  Outcome constructs of criteria identification

The expected outcomes of the criteria identification phase include 1) the selection of comprehensive criteria, and 2) the selection of efficient criteria. Comprehensive criteria denote both the criteria that measure the performance of the health delivery system and so-called "external" indicators providing social characteristics of counties, where the delivery of services takes place. The criteria are efficient if they can help identify the funding allocation (based on the ranking of counties) that has a high potential to result in hiring and longer-term retention of health care professionals. The potential for attracting and retaining a health care professional can be estimated using past data records for the same or geographically similar areas.

### 5.3.4  Criteria identification task – a collaborative approach scenario

In this section we describe a scenario of a collaborative approach to implementing the criteria identification task. The emphasis is on presenting software tools that can be used to support the collaboration. The goal of criteria identification is to select the set of comprehensive and efficient criteria for the evaluation of health care needs in Idaho counties. Based on the task analysis two different meeting arrangements — same place/same time and different place/different time — can be considered for collaborative criteria identification. In the presented scenario the latter arrangement is used.

The participants collaborating on the identification of evaluation criteria include the directors of Idaho's six health care districts, community leaders, hospital administrators, doctors and mid-level providers, and health care analysts from Idaho Department of Health and Welfare. The analysts provide the technical support for stakeholders and decision makers. The technical support covers the range of possible questions about the meaning of proposed criteria, current and expected criterion outcomes in the respective counties, inter-criteria relationships, and the efficacy of criteria in identifying the need for funding. Because of the anticipated delay in providing answers to questions posed by stakeholders and decision-makers, the preferred meeting arrangement for the collaboration is a WWW-based shared workspace system. Many shared workspace systems, such as Lotus Notes, Group Systems or Zeno, could potentially be used. We decided to use the Zeno system (Gordon and Karacapilidis 1997, Gordon *et al.* 1997) because of its highly transparent user interface, argumentation support functions, and the possibility of choosing between a Java applet-based client and a pure HTML client.

Zeno is a groupware tool supporting moderated and non-moderated collaborative work. It was developed by the Mediation Systems team at the German National Research Center for Information Technology (http://ais.gmd.de/MS). The capabilities of Zeno include common groupware tool functions and argumentation support capabilities. The groupware functions of Zeno include document access control, management of user account, storage of addresses and user lists, customizable personal message lists, and support for workflow agendas. The argumentation capabilities of Zeno follow the gIBIS model (Conklin and Begeman 1989). The user can participate in a discussion by posting an *issue* or a *position* concerning an issue. Issues can be commented upon and questioned. Position may be replied to in the form of pro and con arguments. The flow of discussion can be managed by a moderator using three subdirectories: incoming, published, and index (see Figure 5.1). The first receives messages sent by participants in their original state. The second retains the messages approved by the moderator who can change the message's content if it has been determined to be imprecise, change the argument type if it has been classified incorrectly by the user (e.g. from position to comment), or stop the message if it is deemed offensive to another participant. In the non-moderated discussion the incoming messages are automatically published. The index subdirectory provides a structured, tree-like view of the flow of discussion (Rinner 1999).

In the presented scenario, the collaboration on selecting the evaluation criteria takes the form of moderated discussion in Zeno. The discussed issues concern criteria for evaluating the health service funding needs of counties in Idaho. Figure 5.2 presents a structured view of the discussion flow. The first issue "Availability of On Call Providers" concerns the weekly number of on-call hours for each primary health care provider, and is a candidate evaluation criterion. Higher than the average number of on-call hours is usually associated with higher demand for health care services and may serve as an indicator (evaluation criterion) of need for additional

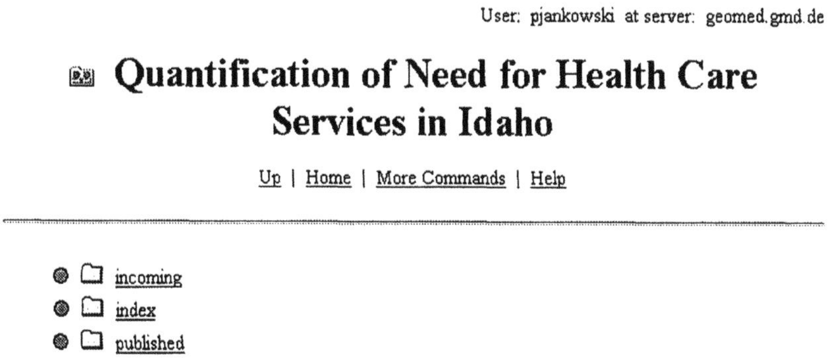

User: pjankowski at server: geomed.gmd.de

## ▣ Quantification of Need for Health Care Services in Idaho

Up | Home | More Commands | Help

⬤ ☐ incoming
⬤ ☐ index
⬤ ☐ published

*Figure 5.1* Discussion forum in Zeno contains three default subdirectories: "incoming", "index", and "published". The discussion forum in this case concerns quantification of needs for health care services in Idaho

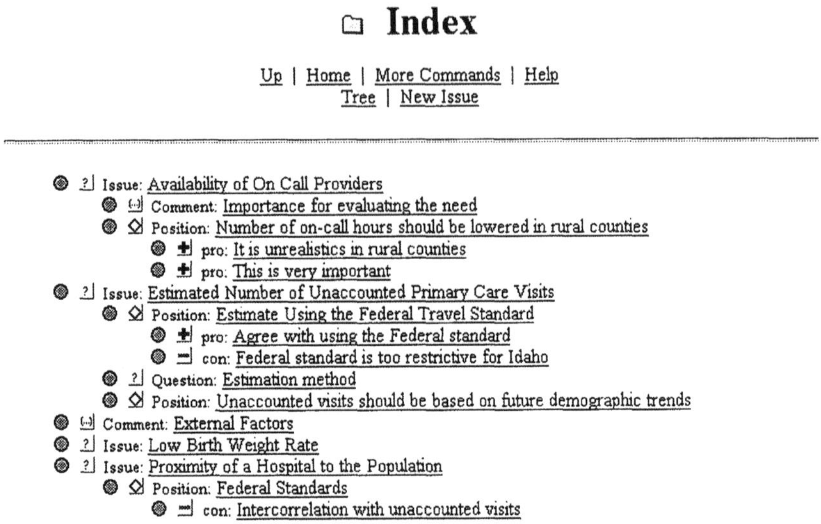

## ◻ **Index**

Up | Home | More Commands | Help
Tree | New Issue

---

⊚ ?⌋ Issue: Availability of On Call Providers
    ⊚ ⌐⌐ Comment: Importance for evaluating the need
    ⊚ ⊘ Position: Number of on-call hours should be lowered in rural counties
        ⊚ ✚ pro: It is unrealistics in rural counties
        ⊚ ✚ pro: This is very important
⊚ ?⌋ Issue: Estimated Number of Unaccounted Primary Care Visits
    ⊚ ⊘ Position: Estimate Using the Federal Travel Standard
        ⊚ ✚ pro: Agree with using the Federal standard
        ⊚ ⌐ con: Federal standard is too restrictive for Idaho
      ⊚ ?⌋ Question: Estimation method
      ⊚ ⊘ Position: Unaccounted visits should be based on future demographic trends
⊚ ⌐⌐ Comment: External Factors
⊚ ?⌋ Issue: Low Birth Weight Rate
⊚ ?⌋ Issue: Proximity of a Hospital to the Population
    ⊚ ⊘ Position: Federal Standards
        ⊚ ⌐ con: Intercorrelation with unaccounted visits

*Figure 5.2* Content of the index subdirectory with the discussion flow tree

resources. The participant who submitted this criterion attached an explanation for the issue, which once published can be accessed by all participants (see Figure 5.3). An explanation can be attached to any type of argument supported by Zeno. In the lack of an explanation, the participants may post questions regarding the meaning of arguments (see question in Figure 5.3 about the estimation method used to compute the number of unaccounted primary care visits).

Besides being questioned, the submitted issues can be commented upon. The other type of argument made about an issue is a position, which in turn can attract a supporting argument (pro) or a critical argument (con). In the current version of Zeno the moderator performs the evaluation of discussion and the discussion summary. In the presented scenario the evaluation of discussion is performed by considering all arguments made about each submitted issue. If the majority of arguments are in favor the issue becomes an evaluation criterion. The results of moderated discussion about the evaluation criteria for primary health care need in Idaho counties are summarized in Table 5.1.

### 5.4 Option evaluation

Just like criteria identification, the option evaluation task can be analyzed in respect to its convening, decision process, and decision outcome constructs. Option evaluation in the framework of spatial multiple criteria decision making involves the comparison of locations based on their selected attributes — evaluation criteria. Depending on the type of

⟨?⟩ **Issue:** # Availability of On Call Providers

Up | Home | Reply | More Commands | Help

Author: Piotr Jankowski
Organization:
Date: 17-Sep-99 5:39:44 PM

The criterion is expressed by the number of hours on call for each provider. The number of hours on call is multiplied by the provider's FTE at each location. The results are summed for each PCSA and the total is divided by the overall FTEs in each PCSA. This yields the measure of the average call burden per FTE in each PCSA

**Context:**

  ⊚ ⟨?⟩ Issue: Availability of On Call Providers
    ⊚ ⟨–⟩ Comment: Importance for evaluating the need
    ⊚ ⟨Q⟩ Position: Number of on-call hours should be lowered in rural counties
  ⊚ ⟨?⟩ Issue: Estimated Number of Unaccounted Primary Care Visits
  ⊚ ⟨–⟩ Comment: External Factors
  ⊚ ⟨?⟩ Issue: Low Birth Weight Rate
  ⊚ ⟨?⟩ Issue: Proximity of a Hospital to the Population

*Figure 5.3* Every issue/argument can have an explanation attached. In this example the attached text explains the meaning of the submitted issue "Availability of On Call Providers"

decision problem at hand, option evaluation may involve the classification of locations into decision-relevant classes, selection of the best location, or ranking of locations from the most- to the least-achieving decision objective(s). The last case is pertinent to improving primary health care services in Idaho counties through a targeted funding allocation. The option evaluation can be accomplished by comparing Idaho counties on the bases of adopted evaluation criteria and arriving at a ranking of counties from the most in need of improving primary health care services to the least in need.

### 5.4.1 Convening constructs of option evaluation

The convening constructs of option evaluation task include outcome goal and objective, process goal and objective, participants involved, and decision support tools.

The outcome goal of option evaluation is to rank Idaho counties on the basis of need for health care services. It is expected that such a ranking will provide the basis for the funding decision and ultimately contribute to improving the access to primary health care services. In practical terms this means that more state residents than currently should be able to find a health care provider within 15 miles from their residence. The travel distance of 15 miles from the point of residence to the point of service corresponds to 30-minute travel time, which is the federally accepted standard

*Table 5.1* Evaluation criteria for ranking Idaho counties according to need for primary health care resources

---

*1. Estimated Number of Unaccounted Primary Care Visits.* This criterion serves as an indicator of unmet demand for primary care medical services due to the unavailability of providers within the 15-mile travel distance constraint.

---

*2. Availability of On Call Providers.* Expresses the number of hours on call for each provider.

---

*3. Availability of Obstetrical Care.* Expressed by the number of providers offering obstetrical delivery services in each county.

---

*4. Availability of Emergency Medical Services.* The criterion values are calculated by dividing the sum of ambulances and quick response units by the number of population in each county.

---

*5. Percent of Population Receiving Medicaid/Medicare.* Percentage of population in each county who utilize Medicaid or Medicare medical insurance.

---

*6. Low Birth Weight Rate.* Percentage of low birth weight births per each county averaged from a multi-year interval.

---

*7. Poverty Level of Population.* The criterion values are based on census poverty rates by county.

---

*8. Proximity of a Hospital to the Population.* The criterion values are calculated based on the nationally accepted standard for the maximum travel distance of 35 miles to a hospital in rural areas. The number of individuals residing outside each hospital influence zone (35-mile radius) is calculated for each county.

---

*9. Emergency Room Visits.* The criterion values are based on annually collected hospital emergency room visit data.

---

*10. Unemployment Rate.* Based on most-up-to-date county-wide unemployment rates.

---

*11. Fertility Rate.* County-wide average fertility rates.

---

*12. Language Barrier.* Due to a significant Hispanic community in southern Idaho the language barrier criterion relates to the inability of providers to communicate with Spanish speaking persons residing in the provider's county. The criterion values are estimated by dividing the number of Spanish speaking providers per county by the number of individuals in the county who speak Spanish at home.

---

for adequate access to primary care services. The reality has been that a number of people in Idaho are located farther than 15 miles from the nearest provider (see Plate 5).

A major impediment to improving the access to primary care services has been the insufficient number of primary health care providers, including physicians and midlevel practitioners (nurse practitioners, physician assistants, and nurse midwives), in the rural areas of the state.

Following the outcome goal, the objective is to generate a ranking of counties, based on the demonstrated need for health care services that will provide the basis for the allocation of funds. The funds will be allocated on the need basis, i.e. go to counties that place high in the ranking. The health care need for each county is evaluated according to the 12 criteria listed in Table 5.1.

The ranking of counties is a collaborative decision task organized by Idaho's Center for Vital Statistics and Health Policy and involving the Board of Directors of the Health Professional Loan Repayment Program. The goal of the decision process is an active and equally footed involvement of Board members. The achievement of this goal can be realized through the following objectives:

1   increasing the efficiency of the decision making process; the decision process supported by collaborative spatial decision support software can prove more efficient than a non-supported decision process if an equally acceptable (or better) ranking can be achieved within a shorter time frame;
2   increasing the satisfaction with the decision process;
3   achieving a consensus-based ranking; the ranking of Idaho counties should be as much as possible a consensual solution reflecting the views of the Board majority.

Participants involved in the option evaluation task include stakeholders and decision makers. Technical experts, with the exception of a group facilitator, do not participate directly in the option evaluation phase. They may be called, however, by the stakeholders and decision makers to help explain the relationships among the evaluation criteria. A good understanding of these relationships is essential to specify trade-offs among criterion values and to assign weights to evaluation criteria. Trade-offs and weights reflect values and interests of individual decision makers and are the essential component of the option evaluation task.

Tools used in the option evaluation task include the geographic database of Idaho counties with their attributes representing the evaluation criteria and values representing the outcomes for every criterion across all counties. Other tools include a specialized software for supporting collaborative spatial decision making, computer hardware allowing each participant to access the database individually via a PC connected to the computer network and send input/data to a facilitator's workstation, and a large-size computer monitor display visible by all participants.

### 5.4.2 *Process constructs of option evaluation*

Process constructs of the option evaluation task include the appropriation of tools and task management. The appropriation of tools deals with the question of which decision support tools are likely to be used by participants. The appropriation of tools is conditioned by the organization of option evaluation process (task management). In the decision problem at hand, the task management is carried out by a group facilitator: a person familiar with the software and the problem domain, yet as much as possible impartial to the outcome of the evaluation process. The task is managed as a private/public decision process. This implies that steps of the task alternate between the private and public mode. First, the participants are asked to consider individually (private mode) a final selection of evaluation criteria. The basic tools used to support this step include tabular and chart views of the problem database, map-based visualization tools, and criterion selection list tool. Additional tools may include text or map-based summaries of discussion/argumentation about different dimensions of the problem, and the results of data mining to find strong evaluation criteria ("strong" in the sense of predicting consequences of a funding decision for the retention of health care professionals). After the expiration of time allocated to private-mode criteria consideration, each participant is asked to submit a list of criteria to the facilitator. The submission is carried out in the form of an electronic vote. For this purpose a voting function in the software is invoked. The votes are collected by the facilitator using another software function and their result, including the measure of agreement/disagreement for each voted criterion, is displayed for the entire group (public mode) on a large computer display. Criteria characterized by a high level of disagreement can then be discussed by the group. Depending on the dynamics of the discussion, some controversial criteria may be accepted while others may be dropped or reconsidered.

Next, the facilitator requests that each participant specify his/her relative preferences for the evaluation criteria. The level of preference for an evaluation criterion is considered equivalent with the relative importance of the given evaluation criterion. Information about the relative importance of criteria is represented by criterion weights. The computation of criterion weights can be a tedious process, especially if the number of criteria is large (more than nine criteria are considered as cognitively challenging for decision makers). It is therefore expected that the participants will use criterion-weighting functions available in the software to support weight computation.

Criterion weights are said to be relative, because they account for change in the ranges of variation in criterion values and different levels of importance attached to these ranges (Malczewski 1999). In a situation where decision options (i.e. counties) have a spatial meaning, the relative importance of criteria can also be judged by exploring the distribution of criterion values among the decision options and the resulting spatial patterns. Although no formal techniques for calculating weights based on spatial distribution of criterion values have been developed yet, combining data scatterplots with dynamic maps (Andrienko and Andrienko 1999a) can be used to support a map-based specification of criterion importance.

Similarly to the criterion selection process, the derivation of criterion weights is carried out in a private/public mode. In the private mode the participants are encouraged by the facilitator to derive a set of criterion weights and submit it in the form of an electronic vote. In the public mode, individual weight sets are coalesced by the facilitator with the help of software and the result is displayed for the group's consideration. It is expected that the group can find a consensus-based set of weights. The possible differences can be addressed by a sensitivity analysis of county rankings.

The weights are used in concert with criterion values to compute a ranking of counties. This step is carried out in the public mode by the facilitator using criterion weights and value aggregation functions available in the software. The ranking orders counties from the highest score to the lowest and can serve as the basis for assigning funds to counties. The ranking results can also be visualized by the participants on a map. The spatial distribution of ranking results and the resulting pattern can be used as an additional source of information in making a final decision.

### 5.4.3 Outcome constructs of option evaluation

The expected outcome constructs of the option evaluation phase include the ranking of counties and the consensual basis for making a decision. The ranking of counties is the result of a collaborative process of evaluating counties on the basis of multiple criteria, criterion values, and criterion preferences. During collaboration the participants alternate between the private and public mode of work, communicating the results of criteria selection and deriving criterion priorities. The facilitator who also helps the participants to cooperate towards arriving at the ranking also coordinates the communication. The collaborative approach provides the basis for deriving a consensus-based ranking of the counties.

### 5.4.4 Option evaluation task – a collaborative approach scenario

Almost the same group of participants as in the criteria selection task is involved in developing the ranking of Idaho counties. The only exception is that health analysts are replaced in this phase of the decision process by a group facilitator. The presence of group facilitator is largely determined by the change of collaboration meeting arrangement from the distributed space and time to the same place and time meeting arrangement. Also, the ranking of counties is a more structured decision phase in the sense of available information about criteria outcomes. The emphasis is no longer on the selection of evaluation criteria, but on the relative importance of criteria represented by criterion weights. In the presented scenario, the process of eliciting weights is supported by collaborative spatial decision software GCP (GeoChoicePerspectives™) (www.geochoice.com). Every participant has access to a PC with the software and the decision problem database. The PCs are connected via a LAN, which also includes a facilitator's PC connected to a large, public monitor display. Each partici-pant has access to two out of three GCP modules: ChoiceExplorer™ and GeoVisual™. The former has a number of functions enabling the evalu-ation of decision options and the sensitivity analysis of evaluation results. The latter is an extension of the popular ArcView® GIS software distrib-uted by Environmental Systems Research Institute (ESRI) of Redlands, California, USA. GeoVisual™ displays information about the spatial dis-tribution of criterion values, background information about decision options, and the distribution of evaluation results. Both modules are integ-rated so that the results of evaluation performed in ChoiceExplorer™ can be immediately displayed on a map in GeoVisual™. Additionally, the participants may use the visualization capabilities of GeoVisual™ to amplify their understanding of spatial relationships among the evaluation criteria within the given decision space.

Below we provide a scenario of using both modules in the context of county ranking decision phase. A participant starts with opening the decision data file in ChoiceExplorer™ containing the abbreviated names of evaluation criteria and criterion weights (see Figure 5.4). Initially, the criterion weights are set by default to be equal, or as equal as possible so that the sum of weights equals 100.

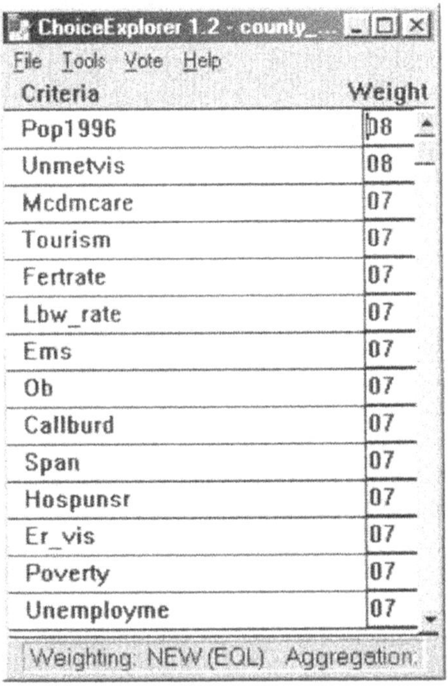

| ChoiceExplorer 1.2 - county_ | _ □ × |
|---|---|
| File  Tools  Vote  Help | |
| **Criteria** | **Weight** |
| Pop1996 | 08 |
| Unmetvis | 08 |
| Mcdmcare | 07 |
| Tourism | 07 |
| Fertrate | 07 |
| Lbw_rate | 07 |
| Ems | 07 |
| Ob | 07 |
| Callburd | 07 |
| Span | 07 |
| Hospunsr | 07 |
| Er_vis | 07 |
| Poverty | 07 |
| Unemployme | 07 |
| Weighting: NEW (EQL)  Aggregation. | |

*Figure 5.4* Decision file containing the names of evaluation criteria and criterion weights in ChoiceExplorer™. Criterion values for each county are not displayed

In order to change the criterion weights the participant can select one of three weight specification techniques: Pairwise Comparison, Ranking and Rating (see Figure 5.5). Each technique relies on a different cognitive mechanism of eliciting preferences for evaluation criteria. Therefore the choice of a specific technique is an individual matter that should be left up to each participant. The Pairwise Comparison technique generates weights based on direct comparisons of one criterion with another criterion (pairwise), in which the participant expresses his/her preference by selecting a rank from a 1–9 scale. The technique assumes that pairwise comparisons are reciprocal. This means that when comparing a pair of criteria (A, B), if the criterion A is preferred to B and it receives the rank of 5, then the criterion B is automatically assigned the rank of 1/5. In the Ranking technique the participant generates criterion weights by assigning a rank from a 1–9 scale to each criterion. Unlike the Pairwise Comparison, the Ranking considers an individual criterion one at a time. In the Rating technique the participant assigns weights explicitly to each criterion by distributing 100 points, which is the sum of all weights.

The specification of criterion preferences can be aided by visualizing the distribution of criterion values. In Figure 5.6, a histobar map of Northern

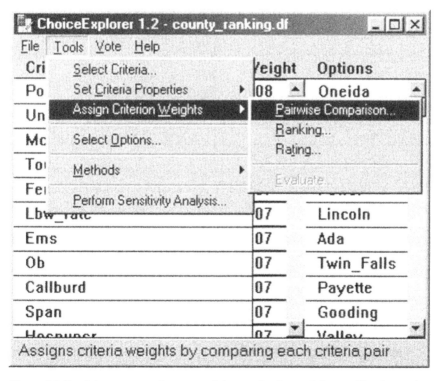

*Figure 5.5* Participants can select one of three criterion weight specification techniques

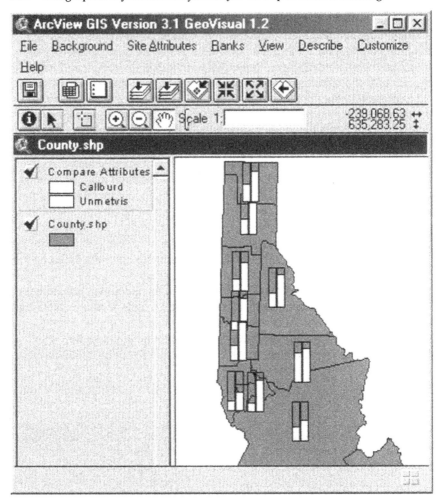

*Figure 5.6* Histobar map in GeoVisual™ presenting the distribution of criterion values

Idaho counties presents a distribution of two criterion values: estimated number of unaccounted primary health care visits (Unmetvis) and the availability of on-call providers (Callburd). The distribution in this part of the state shows that a low availability of on-call providers corresponds to a high number of unaccounted visits. This information may motivate a participant to assign a higher importance rank to unaccounted visits than to availability of on-call providers.

After each participant finishes the specification of weights the group facilitator asks participants to submit their respective weights in the form of an electronic vote (see Figure 5.7). Votes are collected and tallied by the facilitator in ChoicePerspectives™ — the third module of GCP used

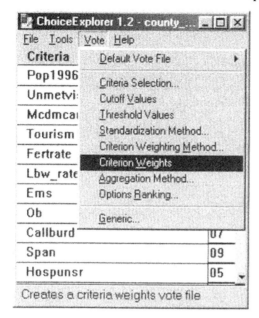

*Figure 5.7* The vote menu in ChoiceExplorer™ allows voting on many aspects of the decision process including criterion weights

for combining multiple perspectives from a group. The purpose of combining votes with presumably different criterion weights is to provide the basis for arriving at a common set of weights. If participants can agree upon a common set of weights (i.e. criterion preferences) then finding a consensus-based ranking of counties becomes much more achievable than in the case of disparate criterion weights. The weights are combined in ChoicePerspectives™ by computing a mean weight value for every criterion (see Figure 5.8).

Results of voting may be discussed and subsequently the participants may agree to adjust the criterion weights. One set of weights is then used to generate in ChoiceExplorer™ a group preferences-based ranking of counties. The ranking can simultaneously be displayed in ChoiceExplorer™ and visualized in GeoVisual™ using the graduated symbol map (see Figure 5.9). The participants may decide at this point to check the stability of obtained rankings using the sensitivity analysis function available in ChoiceExplorer ™. If small changes in criterion weights have no influence on the ranking of options then one may have more confidence in the stability of ranking.

In the described case of evaluating needs of Idaho counties for funding, the results of ranking were used to identify the ten top counties. These were the counties, which in the opinion of participants had the highest need for attracting additional health care professionals. The number of

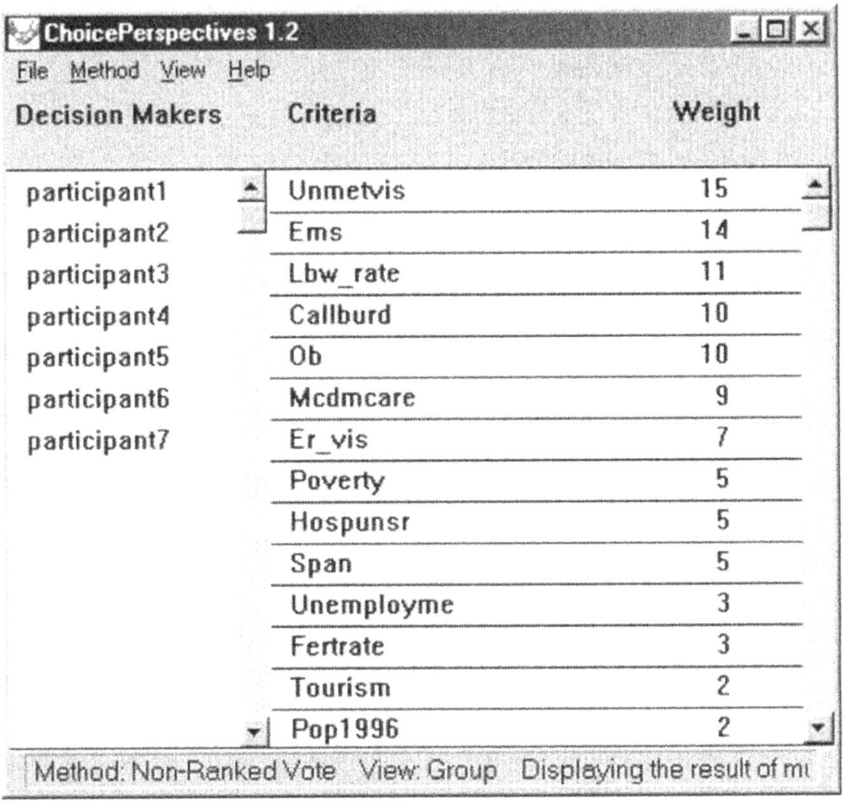

| ChoicePerspectives 1.2 | | _ □ × |
| --- | --- | --- |

File   Method   View   Help

| Decision Makers | Criteria | Weight |
| --- | --- | --- |
| participant1 | Unmetvis | 15 |
| participant2 | Ems | 14 |
| participant3 | Lbw_rate | 11 |
| participant4 | Callburd | 10 |
| participant5 | Ob | 10 |
| participant6 | Mcdmcare | 9 |
| participant7 | Er_vis | 7 |
| | Poverty | 5 |
| | Hospunsr | 5 |
| | Span | 5 |
| | Unemployme | 3 |
| | Fertrate | 3 |
| | Tourism | 2 |
| | Pop1996 | 2 |

Method: Non-Ranked Vote   View: Group   Displaying the result of mu

*Figure 5.8* Mean criterion weight values computed from seven submitted votes

counties was limited by the amount of available funds (approx. US$250,000). Facilities in these counties were then notified that physicians that decided to locate in their area could qualify for loan assistance. The physicians would then have to submit an application to the Department of Health and Welfare for review by the Board of Directors of the Health Professional Loan Repayment Program.

## 5.5  Future potential for enhancing collaborative approach to funding primary health care services

The scenario of collaborative approach to allocating funding for primary health care services presented in sections 5.4.4 and 5.5.5 suffers from the lack of an appropriate method to quantify subjective external factors, such as the financial viability of health care facilities and internal politics of the community. These factors can have a tremendous effect on the availability of health care to special populations such as the poor and migrant and seasonal farm workers.

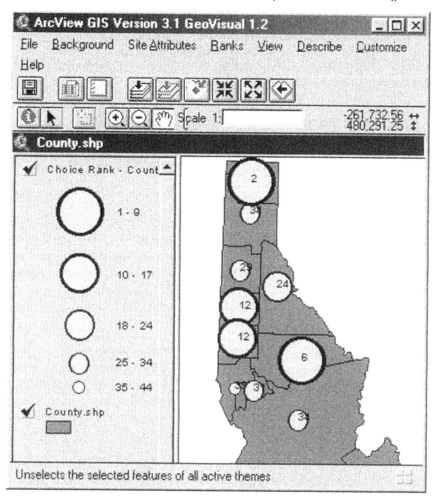

*Figure 5.9* Option rank map presenting ranking results for Northern Idaho counties

The ability of a community to retain a physician is also an important factor. Criteria such as income potential, housing availability, schools systems, and other social factors can help to determine whether or not a community can sustain a provider. These external factors were the subject of heavy debate amongst the board members. Some of the counties that ranked toward the top of the list seemed to be intuitively inappropriate for the Board members. Furthermore, the Board members felt that some of the chosen criteria were contributing little in terms of determining the need of counties for funding. They postulated that new analytical capabilities be built into the system that would allow the group members to check the appropriateness of candidate criteria in terms of their explanatory power. If indeed some criteria could be eliminated without compromising the discriminatory power of the multiple criteria decision model, then the result would be a less complex evaluation process. We address this problem more in depth in the remainder of this section, while demonstrating prototype software.

### 5.5.1 Reducing the cognitive complexity of a multiple criteria spatial decision problem

The cognitive complexity of a multiple criteria spatial decision problem can be considered in both criterion outcome space and geographic decision space. The number of decision (evaluation) criteria can express cognitive complexity in the criterion outcome space. In most of the compensatory multiple criteria decision making (MCDM) techniques the user/decision maker is asked to explicitly express his preferences by assigning numeric weights to decision criteria. The higher the number of criteria, the more cognitively difficult it becomes to consistently assign weights that truly reflect the relative importance of criteria (Jankowski *et al.* 2001). Assigning weights becomes even more confounded through the fact that weights account for changes in the range of values for each evaluation criterion, as well as for varying degrees of importance attached by decision makers to these ranges (Kirkwood 1997). Hence, some compensatory MCDM techniques use the implicit representation of preferences through criterion trade-offs and aspiration levels (Jankowski *et al.* 1999).

The reduction of the cognitive complexity of criterion outcome space can be achieved by means of standard statistical procedures, e.g. multiple regression, discriminant analysis. These procedures, however, require an expert statistician to run them and interpret results, hence they are not attractive for a participatory spatial decision support system involving users with various levels of expertise in analytical methods, such as the case of Board members. Instead of a confirmatory approach, we propose the use of exploratory techniques from the domain of data mining algorithms. One such algorithm promising to be useful in the reduction of criterion outcome space is C4.5 Classification Tree procedure (Quinlan 1993). The goal of this algorithm is to discriminate among some given classes of objects and produce their collective descriptions on the basis of values of attributes associated with class members. In order to achieve this goal, the algorithm tries to select from the available attributes the most discriminating ones. In many cases a small subset of attributes is sufficient to "explain" a given classification. These attributes may be regarded as the most important for characterizing the classes. The capability to select a smaller subset of representative attributes can be exploited for the purposes of complexity reduction.

Cognitive complexity in the geographic decision space is simply expressed by the number of feasible decision options. Additionally, the complexity may be represented by spatial relationships and spatial patterns formed by locations of decision options. Strict numerical representations of these relationships may be difficult to interpret by decision makers, yet they may be capable of eliciting patterns and relationships from maps and integrating them with their heuristic knowledge about the decision situation. Hence, it becomes important in a group decision support system to provide explicit means of expressing this knowledge and including it as a component of the multiple criteria spatial analysis.

Cognitive complexity of geographic decision space can be reduced by applying the concept of Pareto-dominance (Malczewski 1999). A non-dominated decision option, according to Pareto-dominance, is the option that is no worse than the other options on all evaluation criteria and better on at least one criterion. The application of the Pareto-dominance rule results in subdividing the set of decision options into dominated and non-dominated, hence simplifying the structure of geographic decision space.

### 5.5.2 Reducing the dimensionality of health care funding problem

Assigning weights to 12 decision criteria, as in the case of Idaho counties health care funding problem, may be a complex task. In order to reduce the multidimensionality, and hence the complexity of the problem, we developed an integrated dynamic mapping–decision analysis support system called DECADE (**D**ynamic, **E**xploratory **Ca**rtography for **De**cision Support) (Jankowski *et al.* 2001). The system is comprised of the exploratory mapping software Descartes (Andrienko and Andrienko 1999a), data mining software Keppler (Wrobel *et al.* 1996), and an MCDM module. Currently, the link between the systems Descartes and Kepler allows the user to proceed from viewing a map, to application of data mining procedures, to the data depicted in the map. Whenever possible, viewing and interpretation of data mining results obtained in Kepler are supported by means of a dynamic link between graphical displays of Descartes and Kepler, i.e. simultaneous highlighting of corresponding elements in both systems. The architecture of the link between Descartes and Kepler is described in Andrienko and Andrienko (1999b).

The MCDM component of the system is based on Ideal Point technique (Hwang and Yoon 1981). Ideal Point method relies on the notion of the best possible set of criterion outcomes as influenced by three aspects: the Ideal, its Nadir (i.e. worst combination of criterion scores), and the distances from each option to the Ideal and the Nadir. Consequently, in the Ideal Point method each decision option is compared to the Ideal and the Nadir. The criterion outcome distances measured to the Ideal and Nadir, calculated for each decision option, are then aggregated into relative closeness measures. The closeness measure expresses how close each option is to the Ideal and conversely how far it is to the Nadir. The options are then ordered/ranked beginning with the one closest to the Ideal and furthest from the Nadir, and ending with the one furthest from the Ideal and closest to the Nadir. The maximum possible evaluation score (the relative closeness measure) is 100, and the minimum possible is 0.

The user starts the system by selecting decision criteria from the list of candidate criteria and choosing a map display. Three types of map displays for multiple criteria decision analysis are available: *decision options outcome map*, *interactive classification map*, and *criteria outcome map*. The first two are used to reduce the dimensionality of the decision problem and are presented below. The third type of map facilitates the integrated visualization of criterion and decision spaces and is discussed in the subsequent section.

The decision options outcome map interface is organized around a map depicting the geographic extent of the decision situation — in this case the map of Idaho by counties (see Plate 6). On the left of the map the user finds a criterion weight panel providing the listing of decision criteria (ten criteria were selected and, hence, are listed in the panel). Above each criterion name there is a slider allowing the user to select/adjust a criterion weight within the value range (0, 1). The adjustment of one weight causes all other weights to automatically change values proportionally to their values before the adjustment. The north-east pointing arrow, to the left of each weight slider, indicates a benefit criterion (benefit means criterion values are maximized). The user can easily change the directionality of criteria (from maximize to minimize) with one mouse click. The panel to the left of criterion values displays the outcomes for each Idaho county (in this case, decision option) on the plot of parallel coordinates. All three panels — map, criterion weights, and decision option outcomes — are dynamically integrated, in the sense that any action (e.g. option selection, change of weight) in one panel is immediately reflected on the other two panels. Such a level of interactivity allows simultaneous exploration of the decision option outcomes (graph) and geographic decision space (map).

On the screen copy presented in Plate 6, the user selected in the decision option outcome panel 11 counties with the highest final evaluation score. White-segmented lines highlight these counties and the final score can be read from the bottom axis, scaled from 0 to 100. The lines in the decision option outcome panel can be considered as trajectories of decision option outcomes. Their distribution can be easily studied along each criterion axis and give a sense of the influence of that criterion on the final evaluation outcome. Any change of weights causes the final evaluation score to be immediately recomputed and the order of counties along the final score axis changed. This facilitates a sensitivity analysis where the user may select a group of trajectories representing specific counties and test if a slight change in criterion weights causes a change in the final score of selected counties.

The 11 selected counties are also visible on the map of Idaho — their borders are highlighted with white lines — in correspondence with the selected counties on the parallel coordinates plot. The map is dynamically updated in response to changes in criterion weights and final scores. The map design is based on a bi-chromatic color scheme that can be set by the user. In the example discussed here, the bright red was assigned to decision options with high final evaluation score and dark gray to options with the low final score. The boundary of the top ranked county (Washington) is highlighted in yellow and its criterion outcomes are displayed below the map. The corresponding trajectory representing criterion outcomes is also displayed in yellow. Such a selection can be made for any decision option either from the map or from the decision option outcome panel.

Another type of map supporting spatial decision analysis and available in DECADE is the interactive classification map (see Plate 7). The interactive classification map allows the decision maker to group decision options into classes based on the distribution of final scores and heuristics that are not captured by the structure of the decision model. In the presented decision scenario the user assigned the top scoring eight counties to a category "Fundable" and the next two high-scoring counties to a category "Near fundable". The remaining 34 counties were classified as nonfundable.

The classification of Idaho counties presented in the interactive classification map was obtained by using the results of multiple criteria evaluation and a user heuristic. Now, the question becomes whether the same or similar classification could be obtained with fewer evaluation criteria. If so, then the cognitive complexity of the decision problem might be reduced by dropping the irrelevant criteria. Fewer evaluation criteria mean less difficulty in assigning criterion weights and less uncertainty in making a judgement, especially, if among the dropped criteria are the ones requiring the estimation/prediction of criterion outcomes. The proposed solution to the problem of reducing multiple criteria complexity is to submit the interactive classification map with the decision table (containing all initially used evaluation criteria) to the C4.5 decision tree classification algorithm (Quinlan 1993). The algorithm classifies decision options (i.e. counties) by sorting them down the tree from the root to a leaf node, which provides the classification of options. Each node in the tree specifies the test of relevant evaluation criterion, and each branch descending from that node corresponds to one of the possible values for this criterion. A decision option is classified by selecting the most discriminating criterion (criterion that has the highest discriminatory power), dividing the decision options into two branches based on a test condition, finding at each branch the next most discriminating criterion, and repeating this process until no further nodes representing discriminating criteria can be found.

The DECADE software allows the user to run the C4.5 algorithm directly from the interactive classification map. Before a tree can be constructed the user must select the evaluation criteria to be used. The result for Idaho counties and 12 evaluation criteria shows that there are only three relevant criteria for the classification presented in Plate 6: 1) Low birth rate, 2) Population further than 35 miles from the nearest hospital, and 3) Poverty rate (see Figure 5.10). In other words, the classification can be "explained" by three criteria. In practical terms, one would expect to obtain a similar classification using instead of 12 only the three criteria represented in the decision tree — a significant reduction of problem dimensionality. The top node criterion in Figure 5.10 (low birth rate) splits 44 Idaho counties along two branches: 35 counties for which low birth rate was less than or equal to 6.35% of all births, and nine counties that had low birth rate exceeding 6.35%. The former branch leads into a node "Population further than 35 miles from the nearest hospital" and the latter into a node "Poverty rate". Test conditions at these two nodes result then in four branches leading into leaf-nodes. For example, the top-node condition "Low birth rate greater then 6.35%" along with second-level node "Population poverty" classifies nine counties. These counties are then subdivided into a subgroup of two counties belonging to "Non-fundable" class, and seven counties of which six belong to "Fundable" class, and one to "Near-fundable" class (see Figure 5.10).

The distribution of decision options at any node of the tree can be displayed in the decision options outcome map (see Figure 5.11). In the presented example, seven counties of the most right leaf-node (in Figure 5.10) are highlighted with the white boundary line in Figure 5.11.

In order to test the effectiveness of reduced criterion dimensionality users can re-evaluate decision options using exclusively the criteria in the decision tree. In the presented decision scenario we re-evaluated the

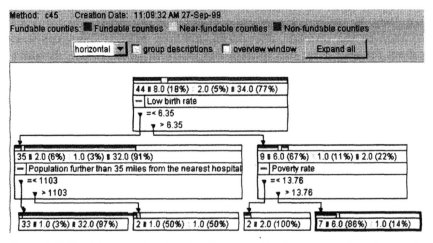

*Figure 5.10* Decision tree for the classification of Idaho counties into three subgroups: fundable, near-fundable, and non-fundable

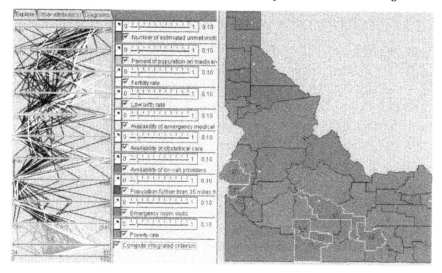

*Figure 5.11* Distribution of decision options at any node in the decision tree can be displayed in the decision options outcome map

funding need of Idaho counties using three criteria contained in the decision tree (Figure 5.12). Next, we compared the result of re-evaluation with the initial evaluation obtained using ten evaluation criteria (see Plate 6). The results of comparison are presented in Plate 8. Altogether 12 counties were selected in two evaluations. Out of 12 counties, eight were selected in the first evaluation (with ten criteria) and in the re-evaluation (with three criteria). This constitutes 67% of agreement between two evaluations. We repeated both evaluations using different sets of weights and obtained very similar results giving us sufficient confidence for reducing the number of evaluation criteria and, hence, simplifying the problem complexity.

In the next step of this exploratory multiple criteria decision making process users could engage in assigning criterion weights and testing the sensitivity of weight changes on final decision option outcomes (see Plate 9).

## 5.6 Conclusion

In this chapter we presented a task analysis-driven approach to participatory spatial decision making, using a primary health care funding allocation problem. The task analysis approach involves identifying and documenting various aspects of a decision situation and, based on the results of the analysis, selecting group decision support tools appropriate for the task. Hence, the objective of this task analysis-based approach to group decision support is to anticipate systematically decision support needs to foster efficient, effective, and equitable decision making.

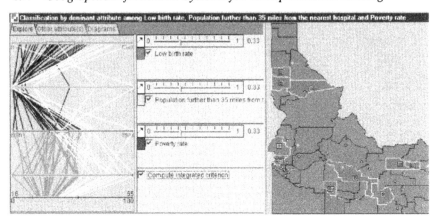

*Figure 5.12*  Top ten scoring counties are highlighted simultaneously in the decision
option outcome panel and on the map. The evaluation was computed
with three criteria

Following the framework presented in Chapter 2, the task analysis of
primary health care funding allocation in Idaho proceeded by examining
convening, decision process, and decision outcome constructs for each task
in the decision situation. The analysis of *convening* constructs allowed us
to articulate what was important in setting up a decision task. The convening constructs included the identification of values, goals, objectives and
criteria shared by the participants and the identification of decision
support tools likely to benefit the collaborative effort. The analysis of
*process* constructs allowed us to consider the dynamics of invoking
decision aids and managing decision tasks. In the case of this analyzed
decision situation, two facilitated modes of participant interaction —
private/public and public — were determined as feasible. The *outcome*
constructs included the selection of comprehensive and efficient evaluation criteria and the consensus-based ranking of Idaho counties on the
basis of the need for primary health care services.

Using the results of task analysis we formulated a scenario for a collaborative approach to implementing two tasks of the decision situation: criteria identification and option evaluation. The emphasis in the scenario
was on presenting group decision support tools that could be used in each
of the tasks. The task analysis of criteria identification revealed that the
participants of collaboration would be likely to have questions concerning
candidate evaluation criteria, which would require a longer period of time
(longer than a single face-to-face meeting would allow) to provide
answers. In view of this we recommended a WWW-based shared workspace system as the preferred meeting arrangement for collaboration.
Zeno was the specific software tool presented in the scenario. We selected
Zeno because of its highly transparent user interface and a rich set of

argumentation support functions. On the example of a moderated discussion we demonstrated then how distributed argumentation software such as Zeno could be used to support the identification of evaluation criteria.

Next, we considered the option evaluation task. Task analysis revealed that option evaluation was more of a structured task than the criteria identification. The emphasis in this task was on developing a ranking of counties using the available information about criterion outcomes for each county and a relative importance of criteria. We found that this collaborative task would lend itself well to the same place and time meeting arrangement, where participants could interact with each other guided by a facilitator and using a GIS-based group support software. For the software to support option evaluation we chose GeoChoicePerspectives. By example of using GeoChoicePerspectives, we demonstrated how spatial decision support software could be used by collaborating participants to evaluate the decision options and run a sensitivity analysis of the evaluation results. We also showed how a GIS component of the software can be used to display information about spatial distributions of criterion values, background information about decision options, and the distribution of evaluation results, and help visualize in a geographic decision space the results of voting on the ranking of decision options.

In the final section we introduced new decision support tools based on the integration of interactive maps, multiple criteria decision models and a data-mining algorithm. The purpose of these tools integrated in prototype software called DECADE is to allow participants to eliminate evaluation criteria that are either redundant or contribute little to the final ranking. Since the elimination of "weak" criteria is performed with exploratory tools that use maps as a primary interface, the tools can be potentially used by a wide group of participants who are familiar with decision situations, but are not necessarily experts with statistical techniques.

## References

Andrienko, G.L. and Andrienko, N.V. (1999a) "Interactive maps for visual data exploration", *International Journal of Geographical Information Science*, 13(4): 355–74.

Andrienko, G.L. and Andrienko, N.V. (1999b) "Knowledge-based visualization to support spatial data mining", in D.J. Hand, J.N. Kok, and M.R. Berthold, (eds) *Advances in Intelligent Data Analysis* (Proceedings of the 3rd International Symposium, IDA-99, Amsterdam), Lecture Notes in Computer Science, vol. 1642, Berlin, Springer-Verlag: 149–60.

Center for Vital Statistics (1994) *Health Needs Assessment for the State of Idaho*, Boise, Idaho, Division of Health, Idaho Department of Health and Welfare.

Chun, K.J. and Park, H. K. (1998) "Examining the conflicting results of GDSS research", *Information and Management*, 33: 313–25.

Conklin, J. and Begeman, M.L. (1989) "gIBIS: a tool for all reasons", *Journal for the American Society of Information Science*, 40(3): 200-13.

Gordon, T.F. and Karacapilidis, N. (1997) "The Zeno argumentation framework", in *Proceedings of the Sixth International Conference on Artificial Intelligence and Law*, ACM: 10–18.

Gordon, T.F., Karacapilidis, N., Voss, H. and Zauke, A. (1997) "Computer-mediated cooperative spatial planning", in H. Timmermans (ed.) *Decision Support Systems in Urban Planning*, London, Spon: 299–309.

Hwang, C.L. and Yoon, K. (1981) *Multiple Attribute Decision Making Methods and Applications: A State of the Art Survey*, Berlin, Springer Verlag.

Jankowski, P. and Ewart, G. (1996) "Spatial decision support systems for health care practitioners: selecting a location of rural health care practice", *Geographical Systems*, 3: 279–99.

Jankowski, P., Lotov, A. and Gusev, D. (1999) "Application of multiple criteria trade-off approach to spatial decision making", in Jean-Claude Thill (ed.) *Spatial Multicriteria Decision Making and Analysis: A Geographic Information Sciences Approach*, Aldershot, Ashgate.

Jankowski, P., Andrienko, G. and Andrienko, N. (2001) "Map-centered exploratory approach to multiple criteria spatial decision making", accepted in *International Journal of Geographical Information Science*, 15(2).

Keeney, R. (1992) *Value-Focused Thinking*, Cambridge, Harvard University Press.

Keeney, R.L., McDaniels, T.L. and Ridge-Cooney, V.L. (1994) *Determining and Structuring Goals and Making Tradeoffs*, Seattle, King County Department of Metropolitan Service (available from University of Washington Libraries).

Kirkwood, C.W. (1997) *Strategic Decision Making: Multiobjective Decision Analysis with Spreadsheets*, Belmont, CA, Duxbury Press.

Malczewski, J. (1999) *GIS and Multicriteria Decision Analysis*, New York, John Wiley.

Quinlan, J.R. (1993) C4.5: *Programs for Machine Learning*, San Mateo, Morgan Kaufmann Publishers.

Renn, O., Webler, T., Rakel, H., Dienel, P. and Johnson, B. (1993) "Public participation in decision making: a three-step procedure", *Policy Sciences*, 26: 189–214.

Rinner, C. (1999) *Argumentative Relations between Geographical Objects in a Hypermedia Planning Support System*, unpublished doctoral dissertation, University of Bonn.

Saaty, T.L. (1990) *Multicriteria Decision Making: The Analytic Hierarchy Process*, Pittsburgh, Expert Choice Inc.

von Winterfeldt, D. (1987) "Value tree analysis: an introduction and an application to offshore oil drilling", in P.R. Kleindorfer and H.C. Kunreuther (eds) *Insuring and Managing Hazardous Risks: From Seveso to Bhopal and Beyond*, Berlin, Springer-Verlag: 349–85.

Wrobel, S., Wettschereck, D., Sommer, E. and Emde, W. (1996) "Extensibility in data mining systems", in *Proceedings of KDD '96 2nd International Conference on Knowledge Discovery and Data Mining*, AAAI Press: 214–19.

# 6    Transportation improvement program decision making

Using proposition analysis in a case study

## Abstract

The Transportation Equity Act for the 21st Century mandates that Metropolitan Planning Organizations coordinate plans, programs and projects within a region. Such coordination is required to be able to receive federal transportation funds for transportation improvement programs. A transportation improvement program is a three-year program of transportation projects that must be created (or updated as the case may be) every two years. The Puget Sound Regional Council prepares the regional transportation improvement program for a four-county area in the central Puget Sound region. In this study we interviewed staff members of the Puget Sound Regional Council to discuss the challenges and opportunities for GIS use as pertains to the regional transportation improvement program. Early on in this study it became readily apparent that there is very limited use of GIS for decision support, i.e. relative to the proposed technology reported by the authors in a previous study related to this topic. Thus, this study became a social-behavioral search about why there has been so little use of geographic information technology when the task is so inherently geographic in character, and GIS technology is readily available. It is not that we suggest that technology should be used, but rather a curiosity of the constraints and/or lack of use. In this chapter we therefore make use of social-behavioral science methodology (outlined in Chapter 4) to explore the character of group decision making, while using Enhanced Adaptive Structuration Theory 2 as the framework. As such, we perform a proposition analysis, as a step beyond construct analysis, as presented in Chapter 5. A construct analysis, based on constructs from Enhanced Adpative Structuration Theory 2 (presented in Chapter 2), followed by a proposition analysis based on premises, is what we call "case analysis". In a case analysis we are in search of explanations about information use and the relationship to decision groups. A construct analysis helps us answer questions about "what", whereas a proposition analysis helps us answer questions about "why". As a report on this case analysis, we perform a construct analysis of the 1999 regional transportation improvement program process. We report on the findings from proposition analysis that takes advantage of the results of a construct analysis. We provide a discussion and interpretation of those findings. A conclusion provides a broader context for what we found.

The Transportation Equity Act for the 21st Century (TEA-21 1998) passed by the US Congress mandates that every metropolitan area in the USA must organize, in cooperation with state transportation organizations, a Metropolitan Planning Organization (MPO) to coordinate plans, programs and projects within a region. Such coordination is required to be able to receive federal transportation funds for transportation improvement. The 1990 (as well as enhancements in 1991) Growth Management Act (GMA), adopted by the Washington State Legislature (Washington State 1990, 1991), mandates that certain contiguous counties growing in population must organize, in cooperation with the Washington State Department of Transportation, a Regional Transportation Planning Organization (RTPO) to coordinate plans, programs and projects across the counties in order to receive state transportation funds for transportation improvement. To avoid duplication, the Puget Sound Regional Council (PSRC) is both the MPO and the RTPO for the central Puget Sound region of Washington State, with lead responsibility to coordinate transportation improvements across King, Kitsap, Pierce and Snohomish Counties (see Figure 6.1).

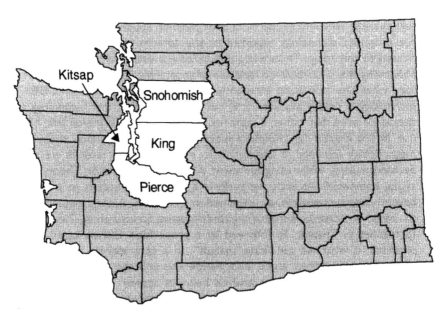

*Figure 6.1* Puget Sound Regional Council four-county region in Washington State

## 6.1 The significance of transportation improvement program decision making

In the USA, a regional Transportation Improvement Program (TIP) is the formal name given to a transportation program document that contains the transportation project list developed through a regional, participatory decision process. A TIP is a three-year program of projects that must be created (or updated as the case may be) every two years. In the 1990s, regional TIPs have been rather different from those in the generations preceding them. The reason is that the Intermodal Surface Transportation and Efficiency Act (ISTEA 1991) passed by the US Congress was a major change in federal transportation law from the 40-year history of the National Defense Highway laws that mandated the US Interstate Highway System. TEA-21 is an updated version of ISTEA, in the sense that it promotes most of the same cooperative policies, but extends and refines them further to recognize that transportation is one among many factors to enhance liveability.

The PSRC (1999a) Call for Regional TEA-21 Projects outlines how governments and other public and private transportation organizations in the central Puget Sound region approach regional transportation project identification, prioritization, selection, and funding. That document is the major driving force in defining the scope and processes for transportation improvement on a regional basis. Appendix A of that document is the Policy Framework for the 1999 TEA-21 TIP Process. In attachment E of Appendix A (p. A18–A19), the PSRC makes it clear to those interested in proposing projects that:

> TEA-21 was passed by [US] Congress and signed into law to continue the intent established under ISTEA to broaden and strengthen the ability of urbanized areas to link their comprehensive planning programs to funding decision on transportation projects. The law states:
> It is in the national interest to encourage and promote the safe and efficient management, operation, and development of surface transportation systems that will serve the mobility needs of people and freight and foster economic growth and development within and through urbanized areas, while minimizing transportation related fuel consumption and air pollution. To accomplish this objective, metropolitan planning organizations in cooperation with state and public transit operators, shall develop transportation plans and programs for urbanized areas of the state. The plans and programs for each metropolitan area shall provide for the development and integrated management and operation of transportation systems and facilities (including pedestrian walkways and bicycle transportation facilities) that will function as an intermodal transportation system for the metropolitan area and as an integral part of the intermodal transportation

system for the State and the United States. The process for developing the plans and programs shall provide for consideration of all modes of transportation and shall be continuing, cooperative, and comprehensive to the degree appropriate, based on the complexity of the transportation problems to be addressed.

In developing the regional Transportation Improvement Program (TIP) to carry out this mandate, TEA-21 requires that each regional/metropolitan planning organization consider at least the following seven factors when preparing its TIP:

1   Support the economic vitality of the metropolitan area, especially by enabling global competitiveness, productivity and efficiency.
2   Increase the safety and security of the transportation system for motorized and non-motorized users.
3   Increase the accessibility and mobility options available to people and for freight.
4   Protect and enhance the environment, promote energy conservation, and improve quality of life.
5   Enhance the integration and connectivity of the transportation system, across and between modes, for people and freight.
6   Promote efficient system management and operation.
7   Emphasize the preservation of the existing transportation system.

Using the above planning factors as a guide, the development of a TIP is a complex group activity because of the number of organizations involved. The PSRC (1999b, p. i) reported on the complex group activity by summarizing:

TEA-21 requires strong decision-making partnerships among local governments, transit agencies, the [Puget Sound] Regional Council, the Washington State Department of Transportation (WSDOT) and other affected public and private parties. The legislation requires the region to develop a [transportation improvement] program that identifies, prioritizes and makes decisions regarding the funding of transportation projects that are consistent with the region's Metropolitan Transportation Plan (MTP).

VISION 2020, which includes the MTP and the local comprehensive plans as required by the [Washington] state Growth Management Act, provides overall guidance for regional and countywide programming activities to allocate regionally managed TEA-21 funds.

Although creation of a TIP is a complex decision activity, the frequency with which they are created, i.e. every two years, provides a basis for policy re-evaluation that helps organize and simplify the process. Such has been the case in the Central Puget Sound region, as the policy frameworks for 1993, 1995 and 1997 developed under ISTEA, were used in succession to help establish the basis for the 1999 TEA-21 TIP Policy Framework. However, it is important to point out that the 1999 Policy Framework under TEA-21 is fundamentally influenced by the seven policy planning factors mentioned above.

In a previous article (Nyerges *et al.* 1998) we performed a multi-organization task analysis for TIP decision making related to the 1997 ISTEA TIP Process, and showed how GIS and group support information technologies can be used by transportation decision groups to enhance consideration of transportation information. We did not examine other aspects of the decision situation in detail, as the purpose was to synthesize task process from three different organizations involved with transportation decision making (among them the PSRC, King County and the Duwamish Coalition) and identify opportunities for technology use based on construction of a normative model of the decision process. The idea was to perform a needs analysis through organizational synthesis, and then propose system requirements that address data management, data analysis, information display, decision analysis and communication management. Considerable potential exists for such technology in TIP decision making.

In this study we interviewed staff members of the PSRC to discuss the challenges and opportunities for GIS use, i.e. scope the issue from the perspective of those who are part of it. Early on in this study it became readily apparent that there is very limited use of GIS for decision support, i.e. relative to the proposed technology reported by Nyerges *et al.* (1998). This study thus became a social-behavioral search about why there has been so little use of geographic information technology when the task is so inherently geographic in character, and GIS technology is readily available. It is not that we suggest that technology should be used, but rather a curiosity of the constraints and/or lack of use. Thus, in this chapter we make use of social-behavioral science methodology (see Chapter 4) to explore the character of group decision making, while using Enhanced Adaptive Structuration Theory 2 (EAST2) as the framework (see Chapter 2). As such, we perform a proposition analysis — as a step beyond construct analysis (let us call it task analysis level 1) of Chapter 5.

A construct analysis, based on constructs from EAST, followed by a proposition analysis, is what we call "case analysis". In a case analysis we are in search of explanations about information use and the relationship to decision groups. A construct analysis helps us answer questions about "what", whereas a proposition analysis helps us answer questions about "why".

As a report on this case analysis, in section 6.2 we perform a construct analysis of the 1999 PSRC-guided TIP process. In section 6.3 we report on the findings from proposition analysis that takes advantage of the results of a construct analysis. In section 6.4 we provide a discussion of those findings to provide an interpretation of what we found. The conclusion in section 6.5 provides a broader context for these findings.

## 6.2 Construct analysis of the 1999 Puget Sound Regional Council TIP situation

The TEA-21 PSRC Call for Projects (Figure in Attachment B, Page A-14) lays out the 1999 TEA-21 TIP Development Process. The tasks and subtasks are as listed in Table 6.1, each considered a macro task step from a macro-micro perspective. Thus, each task in Table 6.1 is the basis for applying the EAST2 framework as depicted in Figure 2.1 of Chapter 2.

For each task or subtask as appropriate as listed in Table 6.1, we coded each of the eight EAST2 constructs in terms of the aspects from EAST2 as they are relevant to that task/subtask. Describing the character of the constructs in this manner is similar to the level 1 task analysis in Chapter 5, i.e. describing the decision situation from macro-task to macro-task. However, as different from Chapter 5, the result of the level 1 task analysis for the TIP decision situation takes the form of a data report that can be used as the "data set" for the proposition analysis (i.e. a level 2 task analysis) performed in the next section. Table 6.2 is an example of a construct coding table, in this case for subtask 1.1. Each of the (sub)tasks becomes an embedded unit of analysis in our TIP decision case study. Yin (1994) argues that embedded units of analysis act as "mini-cases", hence

*Table 6.1* Puget Sound Regional Council 1999 TIP Decision Process

| | |
|---|---|
| Task 1 | PSRC adopts TIP Policy Framework and approves funding allocations for regional and county-wide processes |
| Task 1.1 | Create TIP Policy Framework |
| Task 1.2 | Adopt TIP Policy Framework |
| Task 1.3 | Approve Funding Allocations |
| Task 2 | Regional TIP Project Evaluation Process |
| Task 2.1 | Create and Approve Regional Evaluation Process |
| Task 2.2 | Project Option Generation |
| Task 2.3 | Score Projects |
| Task 2.4 | Initial Evaluation |
| Task 3 | Review and recommend Draft Regional Priorities |
| Task 4 | Conformity analysis on ALL projects and assemble Draft TIP |
| Task 4.1 | Conformity Analysis |
| Task 4.2 | Assemble Draft TIP |
| Task 5 | Public Review and Comment on Draft TIP |
| Task 6 | TPB Recommends TIP action |
| Task 7 | PSRC Executive Board takes final action |

*Table 6.2* Example of construct coding for subtask 1.1

| Task 1.1 Create TIP Policy Framework | 1999 PSRC Transportation Implementation Program Policy framework creation has geographic implications, but no specific locational differentiation until implemented |
|---|---|
| *Convening constructs* | |
| Construct 1: Mandates, objectives, rules, and guidelines | Criteria established by the Transportation Equity Act for the 21st Century (TEA-21) — Public Law 105-178 105th Congress, PSRC charter used for interpretation of law to specify criteria; previous policy used as a guideline for improvement |
| Construct 2: Participants involved | PSRC Transportation Implementation Program personnel |
| Construct 3: Information tools | hardcopy documents shared email exchanged manual document management |
| *Process constructs* | |
| Construct 4: Appropriation of structures | reports of meetings |
| Construct 5: Task management | document reviews face to face meetings |
| Construct 6: Emerging information | suggestions from various responsible parties Appendix 1 of the PSRC (1999a) Call for Regional TEA-21 Projects establish the "regional project evaluation process and criteria". Indeed, this appendix details the construct 2 information that controls what information is emphasized in the decision situation. The appendix material that describes process provides material for detailing constructs 4, 5 and 6 |
| *Outcome constructs* | |
| Construct 7: Decision outcomes | draft policy including criteria associated with policy objectives |
| Construct 8: Social outcomes | systematic overview of organizational processes |

provide a basis for comparison of data observations within a single case study. In effect, they strengthen the evidence base for a case study as they represent the "data protocol" for the case study database which Yin (1994) views as a requirement for rigorous analysis in case studies. The data from the construct coding tables as given by example in Table 6.2 were used to populate the proposition analysis tables described in the next section. (The full complement of construct tables can be found at: http://faculty.washington.edu/nyerges/GISGDM/chap6/construct_analysis _tables.htm.)

## 6.3 Proposition analysis of the 1999 Puget Sound Regional Council TIP decision situation

Propositions are the pair-wise relationships between constructs in a task, hence pair-wise relationships between variables. We use the premises from EAST2 (as in Table 2.5) as the basis for formulating propositions used in this case analysis. We have re-expressed each of the propositions using a pair of research questions to be examined (see Table 6.3). We note that this strategy for research question articulation follows from our discussion in Chapter 2 as to how the propositions in EAST2 "motivate" research questions (see, for example, the discussion relevant to Table 2.5 in Chapter 2).

The research questions listed in Table 6.3 focus the examination of the relationships among constructs. It is these questions that form the basis of the interpretation of the relationship between the paired-constructs. We have chosen to report on two questions for each proposition, although many more questions were posed for possible examination in the formulation of this study. It is up to the researcher to choose what readers might find interesting, based perhaps on what is currently relevant in the extant knowledge of the field. When reviewing the propositions and questions in Table 6.3, note that propositions 5 and 7 from Table 2.5 are not included. Following our discussion of social science research strategies in Chapter 4, the research strategy used in this study does not permit the collection of data to examine those propositions. More micro-scale human-computer interaction data would be needed for investigating proposition 5. Broader-based interview data about social outcomes would be needed to investigate proposition 7.

As mentioned above, the following analyses use the tasks and subtasks of the TIP decision situation as the embedded units of analysis. By comparing the constructs (units) on a pair-wise basis we report on various relationships relevant to the questions being posed. We examined the relationship between grouped pairs of constructs using a table template as a protocol for analysis (see for example Table 6.4). A similar table protocol was used for each question in Table 6.3. The right-most column of the protocol presents the answer for each subtask (embedded units of analysis) relevant to Proposition 1 Question 1. An interpretation of these relationships summarized over the tasks/subtasks, i.e. down the right-hand column, constitute our findings for this research question. All findings in subsequent sections were developed using the same procedure. In the interest of saving space, only Table 6.4 is provided here as an example. (The full complement of proposition analysis tables can be found at: http://faculty.washington.edu/nyerges/GISGDM/chap6/proposition_analysis_tables.htm.)

*Table 6.3* Propositions and questions for focusing case analysis

| Propositions based on EAST2 premises | Research question |
|---|---|
| **Proposition 1:** Basic structures such as mandates, rules and guidelines (construct 1) influence appropriation of information structures (construct 4) | **Question 1:** How do basic structures motivate geographic concerns for transportation improvement decision making? **Question 2:** How do basic structures motivate the use of geographic information structures in decision activity? |
| **Proposition 2:** Participants' perspectives (construct 2) influence appropriation of structures (construct 4) | **Question 1:** Who are the participants and what can be said about their perspectives in relation to appropriation of information structures? **Question 2:** How is each of the perspectives put to use in the decision process? |
| **Proposition 3:** Availability and spirit of geographic information structures (construct 3) influence appropriation of such structures (construct 4) | **Question 1:** In what tasks and by what groups are geographic information structures appropriated? **Question 2:** Why is there little appropriation of geographic information technology in the decision process, given that the process has such significant geographical implications? |
| **Proposition 4:** Appropriation of structures (construct 4), for example, rules and/or maps or tables, influence task management (construct 5) | **Question 1:** In what way do basic structures and/or information structures seem to influence task management? **Question 2:** What decision functions occur in the process? |
| **Proposition 6:** Task management (construct 5) as influenced by appropriation of structures (construct 4) has an influence on decision outcomes (construct 7) | **Question 1:** In what manner are decision outcomes influenced by task management based on the type of basic structures being appropriated? **Question 2:** In what manner are decision outcomes influenced by task management based on the type of information structures being appropriated? |

### 6.3.1 Findings about motivation for geographic concerns in decision activity

Proposition 1 is about the motivation for decision activity, and in particular geographic concerns associated with appropriating information structures in the decision process (see Table 6.3). In Proposition 1 Question 1 we ask: How do basic structures motivate geographic concerns for transportation improvement decision making? Basic structures such as laws, rules, guidelines and policies for an organization are the foundational motivations for decision action. We would expect that these basic structures guide the types of information to be generated.

The federal Transportation Equity Act for the 21st Century and the Washington State Growth Management Act of 1991 provide the primary

Table 6.4 Analysis protocol for Proposition 1 Question 1

| Tasks investigated | Construct 1: structures, e.g. mandates, goals, rules and guidelines | | | | Construct 4: Appropriated structures | Q1 interpretation of relationship between Constructs 1 and 4 |
|---|---|---|---|---|---|---|
| | Name of structure | Type of structure | Structure purpose | Who uses | | |
| Task 1.1 Create TIP Policy Framework | TEA-21 encourages regional TIP | task goal | law mandates 1999 regional TIP | PSRC TIP staff | rationale for 1999 PSRC TIP comes from TEA-21 and growth management laws | TEA-21 is federal transportation law; objectives from TEA-21 are embedded in the policy framework; TIP policy is meant to sustain and improve transportation system as "transportation equity"; hence geographic equity |
| | 1997 TIP Policy Framework | policy guideline | background material | PSRC TIP staff | technical materials for assembling the policy framework; similar process, but different criterion | reinforce the intermodal character of transportation, which tends to be geographic |
| | Growth Management Law | law | guide regional growth | PSRC TIP staff | elements of law encouraging regional transportation system | tendency to focus efforts on central places; rather than wide-spread investments in Regional Transportation Plan |
| Task 1.2 Adopt TIP Policy Framework | draft 1999 TEA-21 TIP Policy Framework | policy guideline | federal and state law | technical analyst; policy makers | fundamental objectives articulated | implementing the policy framework has an inherent geographic bias |
| Task 1.3 Approve Funding Allocations | TEA-21 Guidelines; Growth Mgmt Guidelines | policy guideline | TIP budget constraints | PSRC TIP staff | – previous year funding split<br>– documentation on funds available | history of budget allocation did matter; population road miles is a major concern |

| | | | steps needed to structure the process | PSRC TIP staff; and all others | graphic flowchart | fair (consistent) process was sought – no geographical biases *per se* |
|---|---|---|---|---|---|---|
| Task 2.1 Create and approve Regional Evaluation Process | 1997 Process; Complaints from 1997; 1999 Policy Framework | technical guideline | steps needed to structure the process | PSRC TIP staff; and all others | graphic flowchart | fair (consistent) process was sought – no geographical biases *per se* |
| Task 2.2 Project Option Generation | 1999 Call for Projects, that includes the policy framework | application guidelines | orient jurisdictional proposers | local jurisdiction staffs | local maps — mixed formats; and not necessarily georeferenced with a coordinate system | TIP project application requires a map, but projects are on a variety of linear reference systems, hence do not easily allow compilation of all projects on a single map |
| Task 2.3 Score Projects | 1999 call for projects includes Policy Framework and outline of competitive process | policy and technical guideline | score projects in consistent manner; objectives and criteria established | PSRC staff | matrix scores project database spreadsheet to manipulate weighted scores; call for projects acts as a guide | increase in air quality weights was encouraged from 1997 to 1999. Favors projects that are clean, and conform to emissions regulations |
| Task 2.4 Initial Evaluation | Call for projects includes policy framework based on TEA-21 criteria | policy guideline | objectives and criteria established; projects meet MTP | Regional Project Evaluation Committee | matrix scores | interpreting the weights among objectives impacts the ranking of projects |
| Task 3 Review and Recommend Draft Regional Priorities | 1999 Policy Framework | policy guideline | compare projects in light of MTP and policy framework | Transportation Policy Board | policy framework as a basis for TIP project priority | take into consideration regional needs when comparing projects; regional needs champion over local needs |

*Table 6.4* Continued

| Tasks investigated | Construct 1: structures, e.g. mandates, goals, rules and guidelines | | | | Construct 4: Appropriated structures | Q1 interpretation of relationship between Constructs 1 and 4 |
|---|---|---|---|---|---|---|
| | *Name of structure* | *Type of structure* | *Structure purpose* | *Who uses* | | |
| Task 4.1 Conformity Analysis | Environmental Protection Agency; Clean Air Act Amendment; guidelines for conformity analyses established by Puget Sound Air Quality Control Association | policy guideline; federal regulation | threshold guideline for clean air | PSRC staff | air quality software | three areas of the central Puget Sound region are the focus for conformity analysis: Duwamish Corridor, Kent Valley, Tacoma Tideflats |
| Task 4.2 Assemble Draft TIP | 1999 TIP policy guidelines in call for projects | policy guideline | share priorities with public and decision makers | PSRC staff | priority array | Air quality standards filter out projects all project confirmed |
| Task 5 Public Review and Comment on Draft TIP | TEA-21 encourages public involvement, but does not say what kind | law; policy guideline | open the conversation | TPB and Public | List of projects project maps | unknown if comments had any influence on the TIP project list; availability of citizens in close proximity location at specific times might be favored in public participation |
| Task 6 TPB Recommends TIP Action | 1999 Policy Framework as applied through call for projects | policy guideline | keep process legal | TPB | geographically balanced priority array is intended; all projects must be in line with regional plans which has a definite geographic basis | incorporate the comments and create recommendation |
| Task 7 Executive Board takes Final Action | 1999 Policy Framework as applied through call for projects | policy guideline | keep process legal | PSRC Executive Board | projects with regional implications are part of the priority array; all projects must be in line with regional plans | PSRC Executive Board acts on recommendation in a timely manner |

motivating factors for the PSRC to convene the TIP decision situation. That motivation stems from funds being made available for "transportation improvement". Because transportation projects are likely to be distributed across a region, the law is a major motivating factor in creating geographic information about such projects. PSRC's TEA-21 Call for Projects (PSRC 1999a, page I-1) indicates that there are three major concerns for the magnitude of impacts of any given project:

1 time period of the benefit;
2 who benefits, in terms of many versus few users; and
3 location of the improvement in terms of region-wide versus "spot" improvement.

Each of these concerns is the basis of "scoring" projects prior to project evaluation. As "locational impact" is one of the three major concerns in formulating the TIP, the TEA-21 law thus mandates geographic implications in its authority to allocate funds. Although the TEA-21 law motivates the convening of a TIP process, previous TIP policy frameworks, particularly the 1997 framework, are used to guide the details of the process. Thus, a federal basic structure begets a regional-local basic structure, contextualized to the needs of the central Puget Sound region.

Although the policy framework is motivated principally by TEA-21 and the Growth Management Act (GMA) of Washington State, the federal Clean Air Act Amendment and the Clean Air Washington Act serve as "conforming" basic structures for the TIP, requiring that the TIP maintain or meet National Ambient Air Quality Standards. "Transportation conformity is a mechanism for ensuring that transportation activities — plans, programs and projects — are reviewed and evaluated for their impacts on air quality prior to funding or project approval" (PSRC 1999b, p. viii). Three nonattainment (heavy industry) areas in the central Puget Sound region are the focus of the conformity analyses: Duwamish Corridor, Kent Valley, Tacoma Tideflats.

Together, the mandating and conforming basic structures lay the ground work for institutionalizing the principal, place-based guiding structure, i.e. Policy Framework for the 1999 TEA-21 TIP Process (PSRC 1999a). It is this policy framework that guides the consideration of all information, including geographic information, during the 1999 TIP process. The 1999 Policy Framework based on its predecessors in 1993, 1995 and 1997, has become more clearly focused on "regional priorities" in line with the VISION 2020 30-year (1990–2020) regional transportation plan. "Recognizing the wide gap between total transportation system needs for all levels and modes of transportation and the reality of having significantly less financial resources to meet those needs, ... nine policy objectives have been developed as a guide to help identify and focus the region's near-term investment and funding decisions" (PSRC 1999a, p. A-5).

Consequently, it is clear that projects with regional impacts will be favored in the TIP, and those projects are to be supportive of the region's transportation plans that take the form of a set of maps with descriptions.

In Proposition 1 Question 2 we ask: How do basic structures motivate the use of geographic information structures in decision activity? The TEA-21, GMA, and the 1999 Policy Framework provide the motivation to consider geographic information, but the concern for transportation is itself inherently a geographic problem. As such, these factors motivate the creation of "geographic information structures" as ways of treating information that can be used in a dialog to evaluate priorities for transportation projects. However, it is the Metropolitan Transportation Plan (MTP) mandated by TEA-21 and the Regional Transportation Plan (RTP) mandated by GMA that guide the development of regional project proposals, as each proposal must be a contribution to the overall regional plan if it is to be a serious contender for funding.

In the Call for Regional TEA-21 Projects, an organization must submit a project application that acts as a template for basic structure, and dictates the information structure to be used for proposed projects. Each project application must include a wide array of information that is used to score six criteria: accessibility, urban form, congestion management/system performance, safety and security, air quality benefits/energy savings, and action ready (ability to obligate funds to the project immediately). The ratings for each criterion are assessed by the Regional Project Evaluation Committee in terms of defined impacts that range across low, medium and high. Each of the ratings is defined as appropriate for the criteria under consideration. In general, as mentioned previously, the impact ratings, low, medium, high, for each criterion are defined in terms of near-term benefits, who benefits in terms of many or few, and local versus regional geographic impact. In addition to information that is scored, each project application must include a map of the location of the project with reference to nearest major landmarks. The project information and the associated scores are maintained in a database by the PSRC staff.

Clearly, all of the impacts have a geographic implication. However, it is a bit surprising that the geographic distribution of these implications are not presented to evaluators or decision makers in a geographic manner. Although the TIP application requires a map, the application does not provide a detailed guideline concerning the spatial reference framework to which that map should be related. Consequently, without a standard reference frame, the maps that were submitted tended to use a wide range of referencing systems. They could not easily be compiled onto a single reference framework. Creating a single geographic display with all projects requires considerable effort and knowledge of coordinate systems.

### 6.3.2 Findings about participants and their perspectives on transportation decision activity

Proposition 2 is about "who" participates in the TIP decision activity, i.e. who is invited to "the table" and how the nature of participation influences the decision activity. In Proposition 2 Question 1 we ask: Who are the participants and what can be said about their perspectives in relation to appropriation of information structures?

Examining the nature of participants across the subtasks shows that a wide array of representatives from many agencies take part in the process. Participants are grouped by organization, as they are too many to list individually. The participants consist of people grouped by the following affiliations with the associated responsibilities in the following organizations:

— *PSRC technical staff*, responsible for assembling technical details for policy creation and for project scoring;
— *Regional Project Evaluation Committee* (RPEC), a mix of technical and managerial staff from member jurisdictions, provides a balanced perspective on the initial evaluation, responsible for initial technical and policy evaluation to work through some of the information to make sure it is in a form that decision making body can understand;
— *Transportation Policy Board* (TPB), a mix of elected and appointed officials, responsible for recommending the policy framework and recommending final TIP;
— *PSRC Executive Board*, elected officials of local jurisdictions, elected to the PSRC board by the General Assembly, responsible for approving the policy framework and approving the final TIP;
— *Public*, citizens and affected public and private local organizations.

The groups are responsible for information generation and information evaluation tasks. Technical specialists and managers create details for information consideration. High level managers and elected officials evaluate the information in the context of their responsibility and make decisions. Thus, a stereotypic understanding that "analysts and or managers create information and pass it to decision makers having broader responsibility" is generally observed in this situation. In addition, managerial decision makers pass evaluations on to the elected officials (who have final responsibility) for further evaluation.

The number of participant groups is so large that at times PSRC describes other government agencies or public and private organizations as "the public". Such a reference is not that common when one agency is involved in decision making, as most authors think of individual citizens as the public. However, such a view is in line with Sandman's (1993) research about risk-oriented topics in local communities, whereby he describes nine "types of publics", some of which are government agencies and private organizations. In thinking about similar risk-related decision concerns, other researchers (Rejeski 1993, Stern and Fineberg 1996) place people into three categories according to general roles: technical specialists, who establish alternatives; decision makers, who set policy and evaluate alternatives; and public, composed of interested and affected parties who are impacted by what gets decided. If we were to categorize the various participant groups by role we could say that: the PSRC staff are the technical specialists, the RPEC, TPB, PSRC Exec Board are the decision makers, and the private/public organizations and citizens are the publics who proposed and commented on project lists.

In Proposition 2 Question 2 we ask: How is each of the perspectives put to use in the decision process? Information structures in one task are generated and/or evaluated and passed to the next task where it is acted upon. However, we observe that elected officials from the PSRC Executive Board adopted the policy in the beginning of the process, to "set the stage for information consideration". Then, more technically inclined personnel addressed a number of information generation and evaluation tasks, and then finally passed the results back to the PSRC Executive Board. Thus, the highest level of decision making took place at the beginning and then once again at the end, with the details of projects being treated by managers and technical staff in between.

In the same manner, the PSRC technical staff reviewed the project applications, and used the policy document as a guideline for assigning scores. After the assignment of scores for all projects, the list of projects was passed to the RPEC that would evaluate the scores and place the projects in "policy categories", hence weighting the significance of the projects. The prioritized, and hence reduced, list of projects was submitted for air quality conformity analysis. The conforming list was then submitted for public review.

Examining the responsibility of the various groups in terms of their perspectives from task to task gives a slightly different appreciation of the process, and highlights how iterative the responsibility sharing can be (see Table 6.5).

*Table 6.5* Dominant group perspectives by task/subtask

| Group perspectives | Task/subtask | | | | | | | | | | | | |
|---|---|---|---|---|---|---|---|---|---|---|---|---|---|
| | *1.1* | *1.2* | *1.3* | *2.1* | *2.2* | *2.3* | *2.4* | *3* | *4.1* | *4.2* | *5* | *6* | *7* |
| technical specialists: PSRC staff | • | | | • | • | • | | | • | • | | | |
| decision makers: RPEC, TPB, PSRC Exec Board | | • | • | | | | • | • | | | | • | • |
| publics: priv/pub organizations, citizens | | | | | • | | | | | | • | | |

• = Predominant perspective

In effect, the public, i.e. citizens, local private and public organizations, twice had an opportunity to participate in the process. The first opportunity was when the project applications were being explained in subtask 2.2. The second opportunity was in task 5 once the TIP priority array was assembled. The first opportunity was more of an educational activity in terms of explanations of the criteria for projects, i.e. how to fill out an application and how applications would be rated. The second opportunity was to provide comment on list of projects as evaluated and proposed by the RPEC.

### 6.3.3 Findings about availability of structures for addressing geographic concerns

Proposition 3 is about the availability and spirit of geographic information structures (construct 3) that influence appropriation of such structures (construct 4). To address this we ask in Proposition 3 Question 1: In what tasks and by what groups are geographic information structures appropriated? The project application submitted to the PSRC requires that a map be included, but that map can be of any kind, including hardcopy format, as long as it provides a geographic context for the project. Consequently, the lack of a single, standard framework makes it difficult to compile projects onto the same reference framework. Despite that difficulty, a significant geographic concern is recognized in the process, since the Call for Projects states that the "magnitude of impacts" relate to: time period, who benefits, and geography (location). Geographic impact is recognized as one of the three fundamental concerns, because the continuum of impact scoring ranges from "regional" to "local", with regional impacts receiving a higher score. The detailed scoring criteria, "accessibility" and "urban form" have direct geographic implications. The geographic data are used in the computation of project scores, but ability to view that potential remains somewhat hidden due to the absence of a single spatial reference framework upon which to compile the projects. Consequently, the projects are listed in numerical rank, and described in text form, when passed from Task 2.3 Project Scoring to Task 2.4 Initial Evaluation (when project scores are first treated in the context of policy concerns). Similar scoring lists and descriptions have been used for the past several TIPs.

In Proposition 3 Question 2 we ask: If the spirit of geographic information structures is to provide synoptic access to information, why is there little appropriation of geographic information technology in the decision process, given that the process has such significant geographical implications? During face-to-face interviews with PSRC staff, they mentioned the value of using maps if they had the ability to do so. PSRC does not currently have the fundamental data management capability to "easily" bring all projects to a single map display. Part of the difficulty with data management referencing appears to be technical and another part institutional. Customizing commercial-off-the-shelf software to address the spatial referencing concerns takes time. One attempt at contracting for the solution to the problem resulted in a partial solution, so there is interest in providing more synoptic, geographic information. We address this issue in more detail in section 6.4.

Concerns about "overall project impacts to transportation system" have not been considered. Addressing such concerns would require a multi-scale analysis. Projects are taken on an individual basis, i.e. no inter-project analysis is performed using computer technology, except for the air quality conformity analysis. If such inter-project transportation analysis were performed, then perhaps more participants would call for more use of geographic information technology. However, the timeframe for information generation and evaluation is quite short, and the detailed analysis is not likely to result in significant information gain.

### 6.3.4 Finding about appropriating structures during the decision process

Proposition 4 is about appropriation of structures (construct 4), for example, rules and/or maps or tables, and how they influence task management (construct 5). In Proposition 4 Question 1 we ask: In what way do basic structures and/or information structures seem to influence task management? Overall, creation of one structure leads to the development of another structure. Thus, each task/subtask results in an information product to guide/motivate the subsequent information product. There are two types of structures, hence influences, to consider: basic (administrative) structures and technical information structures.

The basic structures influence the creation of other basic structures or technical information structures, never the converse. That is, the Policy Framework leads to the creation of the Call for Projects document, and the Call for Projects document leads to creation of the project application process. Having a policy that is well thought out guides the process systematically; everyone knows what tasks are to occur and when. Thus, coordination during the process is clear. For those jurisdictions dissatisfied with the prioritization, an appeals process is provided in the document.

In regards to information structure influence, a credible scoring matrix makes for easier scoring assessment of projects. A credible scoring matrix is one that is used by consensus. Thus, if all application information is completed, then projects are easier to score.

Once projects are scored the scores take on a sense of "technical synthesis". This technical synthesis establishes the priorities in the first round. Once scores are evaluated, and projects are assigned to priority categories, then policy-oriented evaluation is performed. Project evaluation based on policy categories is a "big picture" view of the problem. Through this policy weighting of scores a "priority list" of projects is formed. It is this list that is shared with the public.

Public comments about the priority list are considered by the Transportation Policy Board (TPB). The TPB takes the comments into consideration when devising their recommendation. The recommendations are passed to the Executive Board for final decision action as a formalization of the TIP project array.

In Proposition 4 Question 2 we ask: What decision functions occur in the process, and is there any general regularity to the process? Each task within the overall decision situation can be associated with a type of decision function according to Simon's (1979) classification of functions (intelligence, design, choice, review) for organizational decision making, as shown in Table 6.6.

It is interesting to note the embedded iterative nature of the process as hypothesized in the organizational decision literature by Bhargava, Krishnan, and Whinston (1994). However, the overall process is based on the macro decision structure suggested in Chapter 2.

### 6.3.5 Findings about decision outcomes

Proposition 6 is about the influence that appropriation of structures (construct 4) and subsequent task management (construct 5) have on decision outcomes (construct 7). In Proposition 6 Question 1 we ask: In what manner are decision outcomes influenced by task management based on

*Table 6.6* Functions in decision tasks

| Task/subtask | Decision function(s) |
|---|---|
| Task 1.1 Create TIP Policy Framework | intelligence and design |
| Task 1.2 Adopt TIP Policy Framework | choice |
| Task 1.3 Approve Funding Allocations | intelligence, design and choice |
| Task 2.1 Create and approve Regional Evaluation Process | intelligence, design and choice |
| Task 2.2 Project Option Generation | intelligence |
| Task 2.3 Score Projects | design |
| Task 2.4 Initial Evaluation | choice and review |
| Task 3 Review and Recommend Draft Regional Priorities | choice and review |
| Task 4.1 Conformity Analysis | intelligence, design |
| Task 4.2 Assemble Draft TIP | choice |
| Task 5 Public Review and Comment on Draft TIP | intelligence and review |
| Task 6 TPB Recommends TIP Action | design and review |
| Task 7 Executive Board Take Final Action | choice |

the type of basic structures being appropriated? Every subtask has intermediate outcomes that influence the overall outcomes. Since a somewhat similar TIP process was used in 1993, 1995, and 1997, the experience of generating subtask outcomes in a similar manner showed little surprise. However, every subtask was critical to the outcomes. If a group did not do its job, then the process could not proceed.

The outcomes from technical tasks are used as input for decision makers. Technical assembly of the policy framework leads to revision and adoption of the framework. Technical assembly of the project list leads to revision and adoption of the project list. Project criterion scoring was performed, based on definitions in the call for the projects packet. Initially, 72 projects were proposed as a response to the call for projects. After considering policy objectives and budget constraints 42 projects were identified for funding. The public was given the opportunity to comment in writing and at a public meeting. After consideration of comments by policy makers, the 42 projects were recommended for funding by elected officials.

Through Proposition 6 Question 2 we ask: In what manner are decision outcomes influenced by task management based on the type of information structures being appropriated? During the drafting stage of a basic structure it is treated as an information structure. Once the information structure is adopted it can then become a basic structure, but this will not happen in all cases. For example, the policy framework takes on the status of a basic structure. In addition, a different scoring matrix as an information structure can undoubtedly have an influence on the priority of the projects. Different policy objectives applied to the scores could weight the projects differently and undoubtedly result in different prioritization of transportation projects. The federal Clean Air Act Amendment has a substantial influence on the prioritization, as this criterion was weighted quite heavily (even more than the 1997 TIP). Thus, projects that maintain or upgrade the air are highly favored. However, coding of project scores is based on analysts' "shared, but negotiated understanding of project impacts" according to the application response. The scoring can become further clarified over time as further discussion ensues. The 42 projects recommended by the Transportation Policy Board compose the final list of project priorities, but after several iterations of dialog. The Executive Board (of elected officials) approves the priority list as the final list. The list then goes to the Washington State Department of Transportation and on to the Governor's Office. The information structures from task to task do indeed influence the decision outcomes.

## 6.4 Discussion of findings

Performing a proposition analysis in the form of an embedded unit case study allows us to investigate systematically relationships among variables across the (sub)tasks of the 1999 PSRC-coordinated TIP. Using the tasks and subtasks as the embedded units of analysis for each research question, we were able to unpack relationships at the subtask level leading us to interpret what aspects are related in the way they are and why. The selection of constructs and the relationships among them, as expressed through questions motivated by EAST2 propositions, constrain the findings in this study to those constructs (or embedded variables) and relationships. Despite this constraint, the variety of questions asked permit us to discuss our findings from a number of perspectives.

Proposition 1 is about the motivation for decision activity, and in particular geographic concerns. In regards to Proposition 1 Question 1, we asked how basic structures motivate geographic concerns for transportation improvement decision making. We found that motivation for decision activity progressed from subtask to subtask; what was one group's information product was the next group's basic structure for creating an information product. Thus, the policy framework springs from the TEA-21 and GMA laws that motivate decision activity to which funds are allocated for transportation improvement in the central Puget Sound area. The need for project applications springs from the Policy Framework, establishing how information is to be treated in the applications. Although the policy structures that guide creation of information are non-geographic in their form, they are highly geographic in their content implications.

In regards to Proposition 1 Question 2, we asked how basic structures motivated the use of geographic information structures in decision activity. Without a standard referencing frame, other than the Metropolitan Transportation plan, the maps submitted as part of the application package tended to use a wide range of referencing systems. Consequently, the application projects could not be easily collected together on a single reference framework. The PSRC is well aware of this drawback. They contracted a consultant to examine the potential for a single reference framework and thus have an idea of what might be done. In the transportation field this is a rather common problem, and thus transportation GIS researchers have been looking into the difficulties for the past 10 years. Implementing a solution, however, is often more complex, since organizations face both institutional as well as technical constraints on change.

Proposition 2 is about "who" participates in the TIP decision activity. A considerable number of groups participate in the decision activity. The implications for this are both positive and negative, depending on how one interprets "broad-based" participation. In a democratic society such participation is commended. The more organizations/groups that participate, the broader the consensus. Alternatively, one could say that too many groups participating produce a watered-down approach to improvement. A vice-president of a national corporation, who chairs a community-wide committee on transportation improvements, has recently said: "With no one in charge, no wonder transportation congestion on the Eastside of Lake Washington is still so bad". Some people wonder whether anyone involved has a grasp on all the issues. The issues have to be explained so many times to various participants, it is remarkable that all projects contribute in an effective way to transportation improvement. It was found that certain group perspectives focus on certain subtasks. This provides evidence for the suggestion that three major perspectives exist in decision making: policy, technical, and interested/affected parties. Citizens, as interested/affected parties, must spend a rather large portion of their time keeping up with issues in order to be effective. It is no wonder that some of the PSRC material has referred to public and private local organizations as part of the interested and affected parties' "public values" perspective.

Proposition 3 is about the availability and spirit of geographic information structures. To address the proposition we asked in Question 1 about the tasks and groups that make use of geographic information structures. Although basic policy structures call out the consideration of geographic information, and the PSRC staff scored criterion resulting in an assessment of geographic impacts, few geographic information structures were actually used during the decision process. Although the Metropolitan Transportation Plan (MTP) to be upgraded by selected projects is a map, the array of projects could not be presented on those plans in a single coordinate system. Consequently, the only geographic information structures to be used were the MTP and the maps in the individual project applications. Unfortunately, this provides little organization for comparison of the projects — which is the core concern of the scoring matrix that is developed.

In Question 2 we asked why there was so little appropriation of geographic information technology in the decision process, given that the process has such significant geographic implications. What has worked in the past is still working. The nature of the information structures made available to decision makers has stayed relatively stable for the past six years (for development of three TIPs), despite what seem to be considerable improvements in GIS technology. Perhaps one reason for little change involves "what is thought to be needed" during intensive conversations about project prioritization. With intensive conversation, synoptic devices such as maps are not often necessary, because all involved have a good grasp of the information. However, there are so many participants in so many groups involved, that not everyone is likely to have adequate grasp of all information.

Another major reason, and following from Question 1, involves technical and institutional problems with compiling transportation projects onto a single spatial reference framework for synoptic, map access to all projects. Linear referencing is part of data management. Data management underpins the process of getting information to people. Transportation projects require a "linear reference framework", i.e. reference "along the highway", as well as a standard coordinate reference framework in which the linear reference is embedded (Nyerges 1990). The solution to this lingering GIS technical issue has eluded many organizations because its complexity is partly an institutional problem having to do with adoption of a standard reference framework. Without a standard approach, multiple reference systems will continue to be used. A solution to the problem would undoubtedly enhance the public participation aspect of the transportation decision process, because a single map of all projects could be posted on the World Wide Web to spark public comments about evaluation of the projects. Perhaps it is a matter of a cost-benefit computation to guide a decision of whether or not the map-based decision support is necessary.

A simpler approach would be to use a "point" reference in a standard coordinate reference system, i.e. a place holder if the application cannot provide a spatial description compatible with PSRC's recommended linear reference system. A point-based approach has been used quite effectively as the basis for geographic displays that provide geographic context. For example, in a pilot study involving site selection/evaluation for freight mobility projects in the Freight Action in Seattle-Tacoma (FAST) Corridor in the central Puget Sound region, a student group used a group-based decision support software called Spatial Group Choice to create displays to support the negotiation of project selection. A base map with hot links to project photos helped orient participants rather quickly to various rail-highway grade separation locations (see Figure 6.2). Using several criteria describing each of the projects, the group was able to create a multi-participant ranking of the rail-highway grade separation projects that might best address freight movement near and through the FAST Corridor (see Plate 10).

Maps are about spatial relationships, i.e. the interactions among locations. If a computer analysis of the synergetic potential of projects can not be performed because it is not understood how to do it, or it takes too much time, a visual analysis of such interaction potential might be useful, nonetheless. Without the display such analyses are next to impossible because of the amount of information to be considered. For example, with regard to the criterion called urban form, a map of the score for urban form could lead us to a better understanding of the influence of each project on densification of central places as adopted in the comprehensive plan, rather than influence promoting urban sprawl. The boundaries of central places in the central Puget Sound can be displayed, and each project of a particular type can be displayed in geographic relation to these places. Thus, maps are useful information structures in group decision making when used to portray/represent complex relationships that cannot be easily represented by a single measure. However, if a group is not concerned with such relationships, then maps are not likely to provide any more information than could a list of numbers in a table.

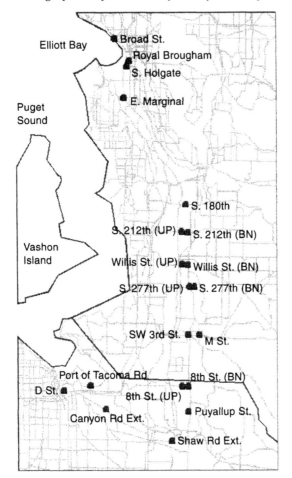

*Figure 6.2* FAST freight mobility (railway grade separation) sites in a decision experiment

Proposition 4 is about appropriation of structures, e.g. rules, maps, and/or tables, and the influence they have on task management. In Question 1 we asked about the way basic structures and/or information structures influence task management. We found that there is a link between the type of structure and the task at hand. Task management seems to be organized around the types of structures adopted to set the stage, and the structures generated to get to the next stage. This is not that surprising since the overall decision situation goal is rather well defined, and the process used to achieve that goal is set out in the policy framework.

In Question 2 we asked about the decision functions that occur in the overall process, i.e. intelligence, design, choice, review. We found that there was a regularity that occurs, as suggested by Simon (1979), but it is an iterative process as hypothesized by Bhargava, Krishnan and Whinston (1994). Such a macro-micro decision process is rather significant as a basis for providing decision support. Despite the fact that multiple groups are participating from step to step in the process, providing decision support tools that emphasize the decision functions — intelligence, design, choice, review — can help groups with their task responsibility in the overall decision situation. Referring back to Table 3.4, tools to support intelligence could involve information management, group collaboration, and display representation capabilities. Tools to support design could involve analysis representation capabilities for configuring potential options. Tools to support choice could involve decision analysis capabilities. Tools to support review could involve judgement refinement and amplification, and analytical reasoning capabilities.

Proposition 6 is about the influence structure appropriation and task management have on decision outcomes. In Question 1 we asked about decision outcomes influenced by task management relative to the type of basic structures being appropriated. Because of the rather large number of organizations participating in the decision process, there were found to be several steps with multiple layers of decision making authority. Final outcomes are undoubtedly influenced by intermediate outcomes. Given that basic policy structures are well defined in form, since the decision process is repeated every two years, the final outcome is not surprising, i.e. a priority list will be created. However, what projects appear on the final list are subject to criterion scoring and policy weighting assessment. "Sensitivity analysis" can check the robustness of the scoring and weighting, but there is no evidence that information tools are available to perform such analysis as part of the basic process followed. Multi-criteria sensitivity analysis tools that create displays, e.g. as used in the pilot study about FAST freight mobility projects in Figures 6.3(a) and (b), can assist with such robustness checks. The effects of small adjustments in weights can be easily detected with such information tools. Although an appeals process is available, the appeals would not normally re-examine the way that all projects are treated relative to each other, since such a case to case examination would be a rather large effort.

In Question 2 we asked about the manner in which decision outcomes are influenced by task management based on the type of information structures being appropriated. As mentioned above, the way that projects are scored can have a significant affect on the placement of that project in the overall list. The weights chosen also have a significant impact on the scores, since the weights are multiplied by raw scores to obtain a weighted score. A score or a weight of "0", when multiplied by any score or weight, will result in "0". The semantic interpretation of the impacts of the project is thus quite important. Initial low score and/or weight assignment will take the project "out of the running" almost immediately, since there are never enough funds to pursue all projects. Having access to information

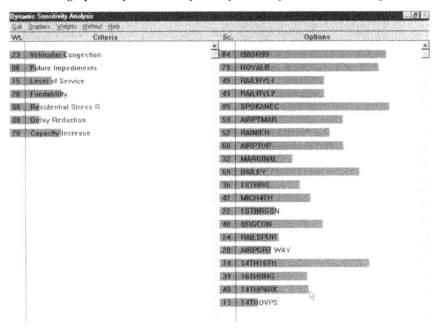

*Figure 6.3(a)* Sensitivity evaluation of project ranking tied to decision table in Figure 6.3(b)

| Criteria | Weight | Options | Score |
|---|---|---|---|
| Vehicular Congestion | 23 | SPOKANEC | 85 |
| Future Impediments | 06 | I90SR99 | 84 |
| Level of Service | 15 | ROYALB | 79 |
| Fundability | 20 | 14TH16TH | 74 |
| Residential Stress R | 08 | BAILEY | 69 |
| Delay Reduction | 08 | AIRPTUP | 60 |
| Capacity Increase | 20 | AIRPTMAR | 59 |
| | | RAINIER | 52 |
| | | RAILRYL2 | 49 |
| | | RAILRYL1 | 49 |
| | | BRGCON | 48 |
| | | MICH4TH | 42 |
| | | 14THPARK | 40 |
| | | 16THBRG | 39 |
| | | 1STBRG | 36 |
| | | MARGINAL | 32 |
| | | RAILSPUR | 24 |
| | | 1STBRGSN | 22 |
| | | AIRPORT WAY | 20 |
| | | 14THOVPS | 12 |

Weighting: USER (EQL)   Aggregation: WSUM

*Figure 6.3(b)* Decision table of project ranking tied to sensitivity analysis in Figure 6.3(a)

tools that create information structures without tedious effort can help ensure that all information is treated fairly. Using several iterations of the sensitivity tools mentioned above can facilitate examination of the stability of the prioritization to alleviate public concern that certain projects are favored unfairly due to "semantic interpretation".

## 6.5 Conclusion

TIP decision making is an institutionalized process in many communities around the world. In the USA, metropolitan transportation decision situations arise every two years, often affecting hundreds of thousands, and sometimes millions, of people. In times of budget austerity, particularly when the current federal transportation law is named "Transportation Equity Act for the 21st Century", and when public involvement is on the rise nationally and internationally, one might think that lots of people would be concerned about the way the transportation improvements come about. Given the number of organizations involved, there are many concerned people. Despite the large number of organizations involved, however, there has been growing concern for the past several years that public involvement is not being facilitated effectively. Public meeting venues conducted in the manner as they have been over the past 30 years since their encouragement by the National Environmental Policy Act of 1969 are implemented in terms of "old social technologies" (CEQ 1997). In decision situations where the risks and stakes are high, the National Research Council has encouraged research about the way that roles (including values and interests) for technical specialists, decision makers, and interested/affected parties come together (Stern and Fineberg 1996). How to encourage information technology improvements to foster public involvement in decision situations is a major concern of many, and one of the fundamental reasons for the study performed in this chapter. To better situate concerns one should have a reasonable knowledge of what the transportation decision situations are about, including knowledge of the major challenges that occur in the process. The finding from any particular study is but one contribution to the overall knowledge building that goes on in research (Brinberg and McGrath 1985). A rigorous socio-behavioral study might lead us to more systemic findings as a contribution to knowledge about decision processes and challenges.

We have situated our examination of TIP project decision making using the EAST2 framework presented in Chapter 2. The propositions of EAST2 express a general relationship about what to expect among the constructs of EAST2. Propositions re-expressed as a set of research questions, elucidate the character of relationships among pair-wise constructs expressed as embedded variables. Using an embedded unit case study has allowed us to investigate systematically relationships among variables across the (sub)tasks of the 1999 PSRC-coordinated TIP. Using the tasks and subtasks as the embedded units of analysis for each research question, we were able to unpack relationships at the subtask level. This systematic reconstruction of the events in TIP decision making suggests to us that rigorous socio-behavioral analyses, such as this one, are needed to make deeper and/or broader contributions to a geographic information science of needs, process, and outcomes in multi-participant settings.

In Chapter 5 we presented a task analysis of the use of group-based decision support software for public health. Such an approach was used on group-based transportation decision making in Nyerges *et al.* (1998). In this chapter we wanted to move to the next level of socio-behavioral rigor by using proposition analysis. Such an analysis has allowed us to explore a number of fundamental research questions, each motivated by fundamental concerns about human-computer-human interaction in a transportation decision situation which faces most, if not all, MPOs around the USA every two years.

After examining the 1999 PSRC-coordinated TIP decision situation, we suggest that the detailed characteristics of policy implementation are more geographic in character than are the general characteristics of the policy, i.e. "geographic transportation reality" is in the details of transportation policy implementation. This suggests that the politics of impacts is more real than the politics of policy, although the latter is more often the conversation among decision makers in governmental organizations. That impact on reality is borne by the interested and affected people who use transportation systems everyday — and not really by appointed representatives who might have people's best interests in mind.

Many groups take part in the TIP decision process over a several month timeframe. The perspective here was inter-organizational participation, like many of the other complex problems our research group has investigated. Despite the variety of perspectives represented, there appears to be continued support for the generalization that technical analysts provide information for decision makers. However, taking this one step further we observed that management groups provide policy information structures for technical work on substantive problem information structures. Those technical information structures are subsequently synthesized into program policy-technical information structures, i.e. the TIP as a package of projects.

Despite the growth and adoption of GIS for technical analysis, there are still fundamental technical-organizational shortcomings that inhibit wider use of the technology, hindering wider distribution of geographic information. The case in point is how to integrate linear referencing systems in transportation in such a way that a synoptic view of all projects can be maintained for everyone throughout a transportation decision process. That includes the public as local public/private organizations, as well as citizen publics who have an interest/stake in what projects are funded for transportation improvement. Technical problems are more complicated when inter-organizational issues, e.g. standards for linear referencing adopted previously by participant organization/jurisdictions in a metropolitan area, are mixed with technical approaches to information management. Standards are social-institutional artifacts of agreement that can be changed, but inertia in institutional processes often retards the change.

We intended this chapter to be a contribution to Participatory GIScience. We view GIS as more than just technology. We view GIS as involving many influences on geographic information use, including technology. Such a perspective is one fundamental difference between GISystems and GIScience. Not finding many GISystems in the TIP process near the beginning of this study raised a question in GIScience: why not, given that the transportation project decision situation is inherently geographic, and there are so many people involved, so that synoptic analysis and display could be rather useful. We believe we have shed some light on the 1999 PSRC-coordinated decision situation in answering that question. Thus, we have contributed to substantive knowledge about GIS use, making use of a theoretically grounded, systematic, qualitative methodology. To really come to know the nature of such decision situations requires further study, as multiple findings are needed to build comprehensive knowledge (Brinberg and McGrath 1985). Extending the case study approach to include multiple cases would provide more evidence about the same research questions (Yin 1994). For example, we are planning (and encourage others) to repeat and/or extend the analysis with previous PSRC-coordinated TIP situations (e.g. the 1995 and 1997 situations), as well those focused at the local government (County and City) level for 1999, and/or other TIP decision coordinating organizations around the world. We encourage others to pursue the same; after all, transportation improvement affects billions of people. The contexts are different, of course, but case study knowledge building can take this into consideration. Proposition analysis can help guide research analysts who want to be more "reconstructively critical" of the use of information products in specific transportation organizations, as well as across decision situations facing communities and society more generally. Reconstructed criticism that conveys suggestions for improvement carries a fuller picture of the situation than criticism alone. We encourage others to join us in this pursuit and let us know what you find.

# References

Bhargava, H.K., Krishnan, R. and Whinston, A.D. (1994) "On integrating collaboration and decision analysis techniques", *Journal of Organizational Computing*, 4(3): 297–316.

Brinberg, D. and McGrath, J.E. (1985) *Validity and the Research Process*, Thousand Oaks, Sage.

Clean Air Act Amendment (CAAA) (1990) US Code, Title 1, Section 103.

CEQ (Council on Environmental Quality) (1997) *The National Environmental Policy Act: A Study of Its Effectiveness After Twenty-five Years*, http://ceq.eh.doe.gov/nepa/nepanet.htm

ISTEA (Intermodal Surface Transportation Efficiency Act) (1991) (PL102–240, 18 December 1991), 105 United States Statutes at Large: 1914–2207.

Nyerges, T. (1990) "Locational referencing and highway segmentation in a geographic information system", *ITE Journal*, March 1990: 27–31.

Nyerges, T., Montejano, R., Oshiro, C. and Dadswell, M. (1998) "Group-based geographic information systems for transportation site selection", *Transportation Research C: Emerging Technologies*, 5(6): 349–69.

PSRC (Puget Sound Regional Council) (1999a) Call for Regional TEA-21 Projects, Puget Sound Regional Council, 5 March 1999, revised 8 March 1999.

PSRC (Puget Sound Regional Council) (1999b) Central Puget Sound Region Transportation Improvement Program Summary, Puget Sound Regional Council.

Rejeski, D. (1993) "GIS and risk: a three-culture problem", in M. Goodchild, B. Parks, and L. Stayaert (eds) *Environmental Modeling with GIS*, New York, Oxford University Press: 318–21.

Sandman, P.M. (1993) *Responding to Community Outrage: Strategies for Effective Risk Communication*, Virginia, Fairfax, American Industrial Hygiene Association.

Simon, H. (1979) "Rational decision making in business organizations", *American Economic Review*, 69: 493–513.

Stern, P.C. and Fineberg, H.V. (eds) (1996) *Understanding Risk: Informing Decisions in a Democratic Society*, Washington, DC, National Academy Press.

TEA-21 (Transportation Equity Act for the 21st Century) (1998) Public Law (PL) 105–78, Title 23 Part 450.324 subpart C.

Washington State (1990) Growth Management Act, Engrossed Substitute House Bill 2929, March 9, 1990.

Washington State (1991) Growth Management Act, Reengrossed Substitute House Bill 1025, June 27, 1991.

Yin, R.K. (1994) *Case Study Research*, 2nd edn, Sage, CA, Thousand Oaks.

# 7  Collaborative decision making about habitat restoration

## A comparative assessment of social-behavioral data analysis strategies

### Abstract

This chapter addresses research questions about the socio-behavioral dynamics of using geographic information system decision aids during collaborative decision making in small, inter-organizational groups. Using an experimental design of a conference room setting, a study of human-computer-human interaction was conducted with 109 volunteer participants formed into 22 groups, each group representing multiple (organizational) stakeholder perspectives. The experiment involved the use of GIS maps integrated with multiple criteria decision models to support group-based decision making. The objective of the decision making activity was the selection of habitat restoration sites in the Duwamish Waterway of Seattle, Washington. Video-taped data were coded using three coding systems: decision functions coding, decision aid coding, and group working relations coding. Although a single set of research questions was used to guide the investigation and hence collect the data, two different types of data analysis strategies were used to process the same data set. We analyzed data from this experiment using traditional statistical inference techniques and exploratory sequential analysis techniques, specifically lag sequential analysis for the latter. We show how different analysis strategies and respective techniques allow researchers to gain information about social-behavioral relationships about human-computer-human interaction from a different perspective. That is, the same research questions, motivated by the conceptual domain, and guided by the substantive issues, are associated with somewhat different relationships in the methodological domain, because the analysis techniques are different. A comparative assessment of the analysis strategies (techniques) shows a difference in information gain. We end the chapter with an evaluation of the appropriateness of different research strategies as suggested in Chapter 4.

In this chapter, premises from EAST2 are used to motivate a set of research questions about the dynamics of group interaction within a collaborative environmental decision making context. An experimental design is used to set up a task situation whereby we can collect fine-grain observations about the use of geographic decision aids. We report how the same data set, based on a single set of research questions, can be analyzed to discover different findings, using two different analysis strategies.

We start the chapter by presenting the premises from EAST2 and the corresponding research questions motivated by these premises, given the substantive context of the study. We then describe the experimental data strategy design used to collect data to examine those research questions. An experimental data strategy design was used in this study because we were interested in the "process interaction" at a fine level of granularity. We adopted a decision task that concerns fish and wildlife habitat redevelopment site selection in an estuary in the City of Seattle Washington. The chosen task is modeled after a real task performed by a multi-stakeholder group convened by the National Oceanic and Atmospheric Administration. For experimental treatments we varied the task in terms of complexity, i.e. fewer sites and fewer criteria, versus more sites and more criteria. Video-tape recordings of the interaction were made and then coded with three coding systems: decision functions that emulate task functions, decision aid structures that describe the software tools, and group working relations that describe conflict. The coding was performed for group interaction at the level of "group attention" with PGIS-based collaborative decision aids. Based on the raw codings we describe the different ways that data were prepared for analysis. We then present our data analysis and the findings associated with each approach, comparing them on a premise by premise, hence research question by research question basis. In the final section we draw conclusions about experimental designs and analysis strategies used to examine human-computer-human interaction during PGIS use.

## 7.1 Posing research questions

We use the premises from EAST2 (as listed in Table 7.1) to motivate research questions to focus our study of collaborative decision making about habitat redevelopment sites. Because we have an interest in the dynamics of human-computer-human interaction during collaborative decision making that makes use of decision support tools, we heeded the research literature to focus on "process" rather than "outcomes" (Rohrbaugh 1989, Todd 1995). Process studies tend to be more meaningful than outcome studies in fine-grained studies of behavior using laboratory experiment designs, because the social-behavioral relations lose "real context". Thus, only the first five premises are addressed, because premises 6 and 7 focus on outcomes. We list a set of research questions that arose, more questions for some premises than others because of our interest.

*Table 7.1* Premises about collaborative decision making and respective research questions focusing on use of geographic decision aids for selection of habitat redevelopment sites

| Premises | Research question motivated by respective premise |
|---|---|
| *Convening premises* | |
| **Premise 1.** Social-institutional influences affect the appropriation of group participant influences and/or social-technical influences | 1.1 How does decision task complexity influence the type of geographic information structures, e.g. maps, tables, diagrams, appropriated by the participants?<br>1.2 How does decision task complexity influence group work? |
| **Premise 2.** Group participant influences affect the appropriation of social-institutional influences and/or social-technical influences | 2.1 Does prior knowledge/experience with decision aids promote more use of maps and decision models?<br>2.2 Does knowledge acquired through group decision making participation promote more effective use of decision models? |
| **Premise 3.** Social-technical influences can affect the appropriation of social-institutional influences and/or group participant influences | 3.1 What is the relationship between map appropriation and decision model appropriation?<br>3.2 What kinds of decision models are appropriated in relation to maps?<br>3.3 Does technology setting have any influence on the type of decision aid use? |
| *Process premises* | |
| **Premise 4.** Appropriation of influences affect the dynamics of social interaction described in terms of group processes | 4.1 What is the relationship between map usage and decison functions, and decision table usage and decision functions?<br>4.2 Are there differences in the level of group conflict associated with different decision phases?<br>4.3 Does task complexity influence group conflict?<br>4.4 Does task complexity influence appropriation during a given phase?<br>4.5 Does group conflict have any influence on appropriation of decision aids? |
| **Premise 5.** Group processes have an affect on the types of influences that emerge during those processes, and emergent influences affect the appropriation of influences | 5.1 Does group work without conflict when examined by task complexity have any influence on decision aid appropriation?<br>5.2 Does group work without conflict when examined by session sequence have any influence on decision aid appropriation? |

Although research questions are motivated by our gaps in knowledge about the use of computer technology during group decision processes, the specific questions articulated in Table 7.1 guide us to adopt certain data strategy design choices. Those choices in turn constrain what kind of data analysis strategy is appropriate. The questions in Table 7.1 were devised in part knowing that our interest was in formulating a laboratory experiment to capture data. We address those issues as part of the discussion of research design in the next section.

## 7.2  Research design

As mentioned in Chapter 4, the phases of research design include choices about treatments, data strategy (social-behavioral setting and data collection instruments) and analysis strategy. However, the core of the design includes choices about socio-behavioral setting, subjects, decision tasks as treatments, and research instruments for data capture, as this core establishes the constraints for data to be analyzed.

The socio-behavioral setting involved groups of five participants assisted by a facilitator/chauffeur using specially developed CSDM groupware software in a decision laboratory. Our choice of five-person groups stems from Vogel's (1993) review of several experiments in GSS research that showed mixed results with groups of three or four, but beneficial results starting with a group size of five. The facilitator/chauffeur provided less mediation than would a facilitator in a large group, and more software (technical) support with the overall problem as would a chauffeur. We simplify our terminology in the remainder of the paper by using "facilitator" to describe this role.

The CSDM software, called Spatial Group Choice, is a research prototype described elsewhere in greater detail (Jankowski *et al.* 1997). Briefly, the software has two modules. The Spatial component is a GIS/mapping module comprised of ArcView 2.1™ software from Environmental Systems Research Institute (ESRI), with a specially configured, simplified user-interface and a set of functionality developed with the aid of ESRI's Avenue™ scripting language. The decision-modeling component, called Group Choice, uses a multicriteria solution for prioritizing alternatives using various weighting and aggregation schemes, together with a group voting capability.

The decision laboratory used for the experiment, located in the Department of Geography at the University of Washington, included six 486 PC stations with seventeen-inch graphics monitors connected to a local area network (LAN). The six PCs were configured in a U-shape layout, so that the five participants and the facilitator/chauffeur (the latter played by the same research assistant throughout the study) could see each other and the public display screen with relative ease. One station was specially configured for use by the facilitator/chauffeur to drive the public display

screen. The Spatial Group Choice software installed on the facili-
tator/chauffeur's station contained several special features to facilitate
group interaction.

The study used 109 participants formed into 22 groups (one group had
only four members). They were recruited from across the University of
Washington campus, and a few from off campus, through announcements
in classes and flyers posted on bulletin boards around campus. No special
competence was sought, only an interest in the environmental decision
task to be undertaken.

Of the 109 participants, 104 finished the study. The average age of the
participants was 28 years. The average education attainment was close to
completion of an undergraduate degree, although there were several
graduate students and participants from off-campus with an interest in
GIS and habitat restoration. With regard to rating "attitude toward
working in groups" on a five point Likert-scale ($-2$ = strong dislike, $0$ =
indifferent, and $2$ = strong like), the participants rated an average of 0.72.
With regard to rating "previous workgroup experience" on a four point
Likert-scale ($0$ = none, $1$ = experiment group only, $2$ = some work
experience, and $3$ = management), the participants rated an average of
2.12. With regard to rating "working with computers on a per week basis"
on a four point Likert-scale ($0$ = none, $1$ = 1–5 hours/week, $2$ = 6–20
hours/week, $3$ = greater than 20 hours/week), the participants rated an
average of 2.02. With regard to rating "experience with using GIS maps"
on a four point Likert-scale ($0$ = do not know, $1$ = heard about it and
tried, $2$ = use some, $3$ = use frequently), the participants rated an average
of 1.21, which indicates overall that the participants were novices. With
regard to rating "experience with using multi-criterion decision models"
on a four point Likert-scale ($0$ = do not know, $1$ = heard about it and
tried, $2$ = use some, $3$ = use frequently), the participants rated an average
of 0.90, which indicates overall that the participants were novices. With
regard to rating "experience with habitat restoration" on a four point
Likert-scale ($0$ = none, $1$ = education only, $2$ = work only, $3$ = education
and work experience), the participants rated an average of 0.95, which
indicates overall the vast majority were novices. Although participants
were novices in both tool use and problem experience, their motivation to
participate was quite high because of their interest in wanting to learn how
to use GIS-based decision support software within the context of an
environmental decision task.

We adopted a realistic decision task to structure our treatments about
site selection for habitat restoration (development) in the Duwamish
Waterway of Seattle, Washington. The decision task was being performed
by the National Oceanic and Atmospheric Administration (NOAA)
Habitat Restoration Panel (NOAA 1993) due to a law suit settled against
the City of Seattle and King County for inappropriate storm sewer drain
management. For years, storm sewer drains had been releasing unfiltered

storm water containing highway gasoline and oil contaminants into Puget Sound (Elliott Bay), degrading the fish and wildlife habitat. A GIS database for site selection problem was compiled from City of Seattle and King County sources. The site selection decision process was expected to involve conflict management during social interaction due to the different perspectives inherent in the views of participating members. Thus, site selection activities are particularly interesting from the standpoint of software tool use and its interplay with group interaction.

Each decision group met for five sessions, one in each of five consecutive weeks (or as close as possible to that schedule), and worked on a different version of the habitat site-selection task. In each of the five sessions per group we asked them to work toward consensus on the selection of three preferred sites (or as many as the $12 million budget would allow) out of the total number of sites presented to them (see Figure 7.1). The total number of sites varied from eight to twenty. At the end of each session, we asked a group to fill out a session questionnaire which provided a means for the individuals to assess group use of the tools, group interaction, and the level of satisfaction with the overall group selection.

We used a counterbalanced repeated measure design for the treatments (Girden 1992). The treatments varied involving: task complexity as the number of sites (eight versus 20), cognitive conflict as the number of criteria (three versus 11 per site), and access to technology (group-and-individual access versus group-only access). With the group-only access to technology, the number of sites was set at 20 and the criteria at 11. Hence, a total of five tasks were constructed. Groups were assigned randomly to start with a task from 1–5 in their first session. They were given a new task sheet in task sequence at the beginning of each of the five sessions, thus cycling through all five tasks by the fifth session. Although we set up the experiment to examine cognitive conflict by varying criteria, we later deduced that this in fact was a change in task complexity as well. Consequently, we did not bother to examine cognitive conflict, but instead looked at task complexity as a variation in both number of sites and number of criteria, with the simplest task involving eight sites and three evaluation criteria per site versus the most complex task being a choice among 20 sites and 11 evaluation criteria per site. In general, we expected that the information technologies would have more positive influences as the task complexity increased.

Data were collected by session (hence task) using questionnaires and coding interaction of videotapes. Each participant filled out a background questionnaire (education, sex, age, etc.) and attended a two-hour CSDM software training session. At that time, we passed out materials introducing the overall habitat site-selection task, assigned the participants to groups based on schedule availability, and handed out stakeholder roles that they could adopt by the time their first decision session convened. Based on interviews completed by the NOAA Restoration Panel (NOAA

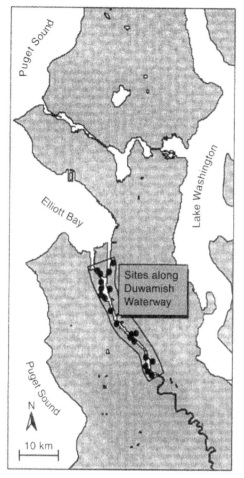

*Figure 7.1* City of Seattle and habitat restoration sites along the Duwamish Water-
       way

1993), participants could adopt a role of business/community leader (20
adopted it), elected official (10), regulatory/resource agency staff member
(22), technical/academic advisor (23), or environmental group representative (29). Roles were self-selected to encourage subjects to participate
based on their inherent interests. We made sure that no less than three different stakeholder roles were represented in each of the groups.

Group interaction was videotaped in each session. A total of 74 sessions
were used for coding, due to group participant attrition within the requirement for five-person decision sessions. Our data sources for the eventual
data analyses are this videotaped record and the session questionnaires.
As we were interested in both software tool use and the overall group
interaction, we focused one video camcorder on the public display screen

and another camcorder (with a remote microphone that hung overhead in the middle of the participants) on the group interaction.

We used a set of interaction coding systems to perform data capture from videotapes on which we recorded the use of CSDM software as a process of group interaction. An interaction coding system is a set of key-words that reliably summarizes the character of a process from a thematic perspective. We make use of three coding systems in this research: a decision aid coding system to describe what decision aids are being used, a decision functions coding system to describe decision activity, and a group working relations coding system to describe group conflict (Nyerges *et al.* 1998).

The decision aid coding system contained four major codes: "MAPS", "TABT", "MCDM", and "CONS". These codes and their subcodes described below were devised specifically for this project based on the software design. Conceptually, the codes represent major differences in "information structuring" that orient a user to information in a particular way. Information structures are the fundamental strategies people use for organizing information portrayal, whether for presentation or analytical purposes (Nyerges 1991).

The MAPS code is an aggregation of the codes for maps implemented with the ArcView 2.1 as the Spatial component of Spatial Group Choice. We used the following codes to describe MAPS available to participants:

BM    (Bar Map) display of site attribute values using bars, as in a bar chart;

CM    (Consensus Rank Map) display of consensus rankings of the sites using circles;

GM    (Graduated Circle Rank Map) display of site ranks using graduated circles;

LM    (Site Location Map) site locations and names only;

OI    (Orthophoto image) shows area using a photo image;

PM    (Previous Rank Map) display of current and previous site ranks using bars;

SM    (Situation/Context Map) situational/contextual characteristics for the sites.

The second major code is "TABT". This code represents codes for data table displays, i.e. displays of data from the data management feature of ArcView 2.1 as part of the Spatial component of Spatial Group Choice. This capability was not coded in any more resolution than "TABT".

The third major code in the decision aid coding system was Multiple Criteria Decision Models (MCDM), representing aids developed as the Choice part of Spatial Group Choice. We used the following codes to describe MCDM aids available to participants:

SC       (Select Criteria) dialog box used to select the criteria, which the group uses for decision modeling;

CV       (Criteria Valuation) dialog box used to value the criteria;

PC       (Pairwise Comparison) weighting method where each criterion is compared against every other for preference;

RK       (Ranking) weighting method which assigns ranks to each criterion a scale from 1 to 9;

RT       (Rating) weighting method with a proportion 100 points across all criteria;

SA       (Select Alternatives) dialog box that reduces the number of alternatives considered in the decision modeling;

MW       (MCDM Window) use of the weights or the decision model evaluation scores and rank list;

SNS       (Sensitivity Analysis Window) use of the sensitivity analysis window.

The fourth major code, "CONS", is an aggregation of the codes used for describing the capabilities available in the Group module of Spatial Group Choice. The CONS code was not subdivided further for analysis since it involved a single table of ranked sites.

The second coding scheme used in the study, the decision function coding scheme, was originally based on work by Poole and Roth (1989a). However, through initial tests of coding we found that those codes did not adequately describe the activity being performed. Thus, we revised the coding system to include the following codes to describe decision function activity:

FUNCS       (Function Structuring) group and/or facilitator activity with a focus on "what's next?", "how do we want to do this?", with respect to which decision function to enter. This would include periods in which the group develops possible approaches to navigate through the decision functions, or to decide which function should be visited next;

PE       (Problem Exploration) periods in which the group attempts to gain a better understanding of the overall problem. The focus of the discussion is on learning about, or investigating, the problem;

CI       (Criteria Identification) group and/or facilitator activity with a focus on identifying, discussing, and selecting criteria that are potentially important for the decision problem;

CV       (Criteria Valuation) group and/or facilitator activity with a focus on selecting the preferred method of valuing the criteria;

CP       (Criteria Prioritization) group and/or facilitator activity with a focus on differentiating the criteria to determine the relative importance of the criteria for the decision problem, or to gain a better understanding of individual priorities;

EA        (Evaluate Alternatives) group and/or facilitator activity with a focus on comparing and contrasting the alternatives;

SA        (Select Alternatives) group and/or Facilitator activity involving statements about the group's actions which explicitly refer to the selection of alternatives, voting, or the reduction of the set of alternatives for the decision selection.

In the third coding scheme used in the study we used two codes to describe group working relations, one that coded group work without conflict (GW) and another that coded group work with conflict (GWC). Group Work Conflict (GWC) were identified as periods when the group is organized and working together, but members disagree with each other or express a different perspective on the topic. This included attempts by group members to manage or diffuse the conflict, or periods when the group is in disagreement because of a misunderstanding. Attempts to diffuse conflict were interpreted from work undertaken by Poole and Roth (1989a) as follows:

- opposition — periods in which disagreements are expressed through the formation of opposing sides;
- accommodation — a mode of resolution of opposition in which one side gives in;
- tabling — a mode of resolution of opposition in which the subject is tabled or dropped;
- negotiation — a mode of resolution in which the group negotiates to manage conflict;
- compromise — a mode of conflict resolution in which the group compromises;
- justification — periods when a supporting rationale for a particular position is posed to the group for consideration.

As was discussed in Chapter 4, data coding can be implemented at a variety of "units of observation" as concerns the granularity of interaction. In this study, coding of data was implemented at the level of "group attention". Group attention is a level of attention at which at least four out of five participants show head-directed awareness of a group-oriented conversation/discussion underway as the predominant activity in a given time cell. A time cell is a one-minute interval (called a count), in which the predominant activity that is occurring is observed by the coder from viewing the video-recording and coding the observation to a database. Thus, for each coding system we created an event/activity sequenced track; observing the videotape three times to code the three (coding system) tracks (decision aids, decision functions, and group working relations) needed to create entries in the database for each videotape session.

To create the database, a SVHS video recorder was connected to a Macintosh computer running the MacSHAPA software (Sanderson *et al.* 1994). MacSHAPA provides an outstanding set of capabilities for sequential data collection from videotape. An interpreted data observation recorded by MacSHAPA and used as the basis of analysis is a coded "time-cell". We worked with fixed-length time cells of one-minute duration, and prepared data for analysis in three ways. One way we prepared data was simply to make use of the time-cell as a one-minute unit of activity, referring to the sum of the cells as "cell counts" or just "counts". A second way was to recode the one-minute activity into "events" lasting for a contiguous number of time-cells — regardless of how many happened to be in a sequence. That is, an event is a contiguous observation of like codes. For example, if a "site location map" was observed to be used in three contiguous time-cells (i.e. three minutes), then the three-minute group of cells can be called an "event" of site location map use. However, rather than refer to the event as the basis of analysis, we refer to a "move" related to an event as the basis of analysis. A "move" is a change from one event into the next event, i.e. when a code changes from one recorded value of any given length to a different value across a cell boundary, then a move exists. The term "move" was adopted from Adaptive Structuration Theory as presented by DeSanctis and Poole (1994). Thus, a move is a start of an event, and can be thought of as the same when it comes to analysis. A third approach to data preparation was the creation of paired-counts or paired-moves. Because a coder could code only one observation (event) in a given time cell, and each coding scheme is composed of several types of activities, we had to recode one-minute cells to two-minute cells to represent a simultaneous use of two decision aids, e.g. decision tables and maps. We discuss this further in the subsection where this coding approach is relevant.

## 7.3  Findings from two types of analyses of the same data

We have used two strategies to perform different types of analysis on the data collected from the experiment. Both analyses make use of quantitative "counts" of coded observations. The purpose for these two analyses was to explore the differences in findings. We expected that the findings would be different, but the real question is: In what way are they different? The first type of analysis follows from a more traditional perspective based on inferential, statistical data analysis techniques. The traditional analysis consists of three techniques: analysis of variance, a general linear model, and a difference of means T-test. The results reported in this chapter are an extract from a more complete analysis reported elsewhere (Jankowski and Nyerges 2001). Here we report on aspects of that analysis in order to compare it with a second type of analysis. The second type of analysis is based on exploratory sequential data analysis (Sanderson and

Fisher 1994). Exploratory sequential data analysis is defined by Sanderson and Fisher (1994, p. 255) as:

> ... any empirical undertaking seeking to analyze systems, environmental, and/or behavioral data (usually recorded) in which the sequential integrity of events has been preserved. The analysis of such data (a) represents a quest for their meaning in relation to some research or design question, (b) is guided methodologically by one or more traditions of practice, and (c) is approached (at least at the outset) in an exploratory mode.

The exploratory sequential data analysis technique we used is called lag sequential analysis, referred to here as sequential analysis.

The premises and research questions listed in Table 7.1 act as the organizing framework for the comparison of analysis strategies. This approach allows us to be specific about the comparison from a theoretical perspective, since the premises are the conceptual grounding for the study. Each of the questions is considered for analysis by the two strategies (see Table 7.2). The questions that each analysis strategy can address are indicated by reference to the technique labeled in the appropriate column of Table 7.2. Where "no analysis" appears in the table, this means that the respective technique was not (and probably cannot be) used to address that question. By reference to the entries in the table, and perhaps not so surprisingly in general, neither of the analysis strategies can address all of the research questions. More surprising is that only four research questions are addressed by both, i.e. questions 1.1, 3.1, 4.1, and 4.2 (these are shown in bold). Consequently, these are the four research questions for which we will discuss the analysis comparison in more detail in the following subsections.

First, we will present the results of the analysis for each strategy. Then, we compare the strategies by making use of the relation feature correspondence criteria introduced in Chapter 4. As previously discussed, the relation feature correspondence criteria consist of 10 features that describe the nature of information gain in an analysis (Brinberg and McGrath 1985). Each of 10 features is treated within a comparison table to provide a sense of the information gain for the respective analysis.

In both analyses, the data (codes) are aggregated to the top level of the coding system to simplify the interpretation of results, although we do lose something in the discriminatory power of the relationships. The decision aid data codes are "MAPS, MCDM, TABT, and CONS". The group working relations codes are "GW and GWC". The decision functions data codes are "PE, CI, CV, CP, EA, and SA".

*Table 7.2* Two types of analyses for the same research questions

| Premise | Research question addressed | Traditional analysis | Sequential analysis |
|---|---|---|---|
| Premise 1. Social– institutional influences | **1.1 How does decision task complexity influence the types of geographic information structures, e.g. maps, tables, diagrams, appropriated by the participants?** | **General Linear Model** | **Lag Sequential Analysis of DFCS to DSCS by aggregated over DFCS codes** |
| | 1.2 How does decision task complexity influence group work? | no analysis | Lag Sequential Analysis |
| Premise 2. Participant influences | 2.1 Does prior knowledge/ experience with decision aids promote more use of maps and decision models? | Analysis of Variance | no analysis |
| | 2.2 Does knowledge acquired through group decision making participation promote more effective use of decision models? | Analysis of Variance | no analysis |
| Premise 3. Information technology influences | **3.1 What is the relationship between map appropriation and decision model appropriation?** | **Pearson Correlation** | **Lag Sequential Analysis of DFCS to DSCS by session sequence** |
| | 3.2 What kinds of decision models are appropriated in relation to maps? | Pearson Correlation: General Linear Model | no analysis |
| | 3.3 Does technology setting have any influence on the type of decision aid use? | no analysis | no analysis |
| Premise 4. Appropriation influences | **4.1 What is the relationship between map usage and decision functions, and decision table usage and decision functions?** | **Analysis of Variance Difference in means (T test)** | **Lag Sequential Analysis of DFCS to DSCS by task sequence** |
| | 4.2 Are there differences in the level of group conflict associated with different decision phases? | Analysis of Variance | no analysis |
| | **4.3 Does task complexity influence group conflict?** | **Analysis of Variance** | **Lag Sequential Analysis of GWRCS, collapsing by DSCS** |

*Table 7.2* Continued

| Premise | Research question addressed | Traditional analysis | Sequential analysis |
|---|---|---|---|
| | 4.4  Does task complexity influence appropriation during a given phase? | no analysis | Lag Sequential Analysis of DFCS to DSCS by phase by task complexity |
| | 4.5  Does group conflict have any influence on appropriation of decision aids? | no analysis | Lag Sequential Analysis of DSCS by GWRCS |
| Premise 5. Emergent influences | 5.1  Does group work with conflict when examined by task complexity have any influence on decision aid appropriation? | no analysis | Lag Sequential Analysis of DSCS by GWRCS by task complexity |
| | 5.2  Does group work without conflict when examined by session sequence have any influence on decision aid appropriaton? | no analysis | Lag Sequential Analysis of DSCS by GWRCS by session sequence |

### 7.3.1  Research question 1.1

How does decision task complexity influence the types of geographic information structures, e.g. maps, tables, diagrams, appropriated by the participants? When undertaking a traditional statistical analysis for this research question we formulated a null hypothesis of the form: Task complexity has no influence on appropriation of map aids (coded as MAPS) and/or decision table aids (coded as MCDM). To determine whether task complexity had any effect on the interaction between the number of map moves and the number of MCDM moves we used a General Linear Model (GLM) statistical procedure (SPSS Base 8.0, 1998). A GLM procedure provides regression analysis and the analysis of variance for one dependent variable by one or more factors and/or variables. The dependent variable was map moves, the covariate was MCDM moves, and the factor was task complexity. Task complexity ranged from task 1 (the least complex: eight sites and three evaluation criteria) to task 4 (the most complex: 20 sites and 11 evaluation criteria). The model explains only 8% of the variability between the use of MAPS aids and MCDM aids (Adjusted R Squared = .081, F = 10.682, Sig. = 0.000), and task complexity is not a significant effect in explaining variability (F = 1.368, Sig. = .252) of the use of these decision aids.

In further examination of the use of MAPS and MCDM decision aids, we were especially interested in the extent to which maps were used concurrently with MCDM aids. However, based on these findings we can conclude that participants used maps more often independently from MCDM aids than in concert with MCDM aids. During the entire experiment (74 sessions), maps were used together with MCDM aids 34.4% of time (conjunctive map use). In the remainder of cases (65.6%) maps were used independently from MCDM aids (disjunctive map use). In order to perform an analysis of "concurrent" use, the data had to be recoded. Coding of decision aid use created only a single entry for each one-minute cell count because the MAPS and MCDM codes were part of the same coding system. Therefore, in order to code concurrent use, one-minute cells were expanded to two-minute cells, and each time a MAPS aid was observed we checked for a MCDM code in the adjacent one-minute cell. Answering the concurrency question was constrained by the implementation of the coding system, as addressed in section 7.3.2.

When considering sequential data analysis, we find that we can address the relationship between task complexity and decision aid use. However, we cannot address the concurrent use of decision aids because of the coding system constraint described above. In that circumstance only a single decision aid code could be used per one-minute count interval. Remember, three coding tracks were used in the coding of the data, however, a track stores only one code in that minute interval. In a lag sequential analysis we code from a base code (F1, e.g. as group working relations in Table 7.3) to a target code (T1, e.g. as decision aids coding system in Table 7.3). The "0" lag is the code that occurs at the same time as the "F1" code. Lags "−5 – −1" are codes that are 5 minutes through 1 minute before the lag "0" code. Lags "1–5" are codes that are 1 minute to 5 minutes after the lag "0" code. For example, in Table 7.3 at the top of the table, when the group work "GW" code appears in the data set, a total of 46 MAP codes were observed, meaning that 46 minutes of map use were coded during 46 minutes of group work, i.e. work with no conflict. The other information in this table is that 44 one-minute observations of map codes exist in the one minute prior to the existence of a "GW" code, and 45 one-minute observations appear after the "GW" code (see the highlighted cell entries in Table 7.3). The amount of information in these lag sequential tables encourages us to simplify the analysis, by focusing on the lag "0".

Reporting on the "0" lag only, i.e. when a code in one coding system (F1) occurs at the same time interval as a code in another coding system (T1), we can create a much simplified table as in Table 7.4. Combining both the "GW" decision aid use together with the "GWC" decision aid use gives us the information in Table 7.4, that is the difference in time for all groups that used decision aids in task 1 and task 4. From now on our findings will be generated using the more simplified version of the tables, i.e. a lag "0" interpretation as in Table 7.4. Interestingly, maps as coded by MAPS were used more in the simple task than they were in the complex task by about twice as much. However, decision tables as coded by

*Table 7.3* Lag sequential analysis from group working relations (F1) to decision aids (T1) as compared between task complexity — task 1: simple and task 4: complex (see text for explanation of codes)

| | | | | | | | | Lags | | | | | | |
|---|---|---|---|---|---|---|---|---|---|---|---|---|---|---|
| | | | −5 | −4 | −3 | −2 | −1 | 0 | 1 | 2 | 3 | 4 | 5 |
| | F1: | T1: | | | | | | | | | | | |
| Task 1 | GW | MAPS | 0 | 0 | 0 | 0 | 44 | 46 | 45 | 0 | 0 | 0 | 0 |
| Task 4 | GW | MAPS | 0 | 0 | 0 | 0 | 15 | 14 | 15 | 0 | 0 | 0 | 0 |
| Task 1 | GW | MCDM | 0 | 0 | 0 | 0 | 46 | 48 | 46 | 0 | 0 | 0 | 0 |
| Task 4 | GW | MCDM | 0 | 0 | 0 | 0 | 73 | 73 | 75 | 0 | 0 | 0 | 0 |
| Task 1 | GW | TABT | 0 | 0 | 0 | 0 | 2 | 3 | 3 | 0 | 0 | 0 | 0 |
| Task 4 | GW | TABT | 0 | 0 | 0 | 0 | 3 | 6 | 6 | 0 | 0 | 0 | 0 |
| Task 1 | GW | CONS | 0 | 0 | 0 | 0 | 9 | 12 | 12 | 0 | 0 | 0 | 0 |
| Task 4 | GW | CONS | 0 | 0 | 0 | 0 | 37 | 44 | 41 | 0 | 0 | 0 | 0 |
| Task 1 | GWC | MAPS | 0 | 0 | 0 | 0 | 11 | 11 | 12 | 0 | 0 | 0 | 0 |
| Task 4 | GWC | MAPS | 0 | 0 | 0 | 0 | 5 | 6 | 6 | 0 | 0 | 0 | 0 |
| Task 1 | GWC | MCDM | 0 | 0 | 0 | 0 | 10 | 10 | 11 | 0 | 0 | 0 | 0 |
| Task 4 | GWC | MCDM | 0 | 0 | 0 | 0 | 52 | 52 | 50 | 0 | 0 | 0 | 0 |
| Task 1 | GWC | TABT | 0 | 0 | 0 | 0 | 0 | 1 | 1 | 0 | 0 | 0 | 0 |
| Task 4 | GWC | TABT | 0 | 0 | 0 | 0 | 3 | 2 | 2 | 0 | 0 | 0 | 0 |
| Task 1 | GWC | CONS | 0 | 0 | 0 | 0 | 17 | 16 | 15 | 0 | 0 | 0 | 0 |
| Task 4 | GWC | CONS | 0 | 0 | 0 | 0 | 21 | 20 | 19 | 0 | 0 | 0 | 0 |

MCDM were used slightly more than twice as much in the complex task than they were in the simple task. Tables to display data were used very little in both tasks. Consensus aids were used a little more than twice as much in the more complex task than in the simple task.

*Discussion of findings and strategy assessment* The way in which the coding was performed constrains both analyses. In regards to the traditional statistical analysis, processing the recoded one-minute intervals using two-minute intervals as if the latter represented simultaneous use of decision aids can be debated successfully both pro and con. Two minutes is so short an interval that most people would agree that use within this interval can constitute "effective simultaneity". After all, why is one minute so special as to be more effective than two minutes? One minute is better because we really should have coded the activity in thirty-second intervals. We missed the opportunity in coding and subsequent analysis when we did not take advantage of the primary and secondary codes in the coding software. That is, MacSHAPA actually permits two codes to be

*Table 7.4* Decision aid use (measured in minutes) distinguished by task complexity

| Decision aid category (DSCS) | Task complexity | |
| --- | --- | --- |
| | Task 1: simple (minutes[1]) | Task 4: complex (minutes[1]) |
| MAPS | 44 | 21 |
| MCDM | 46 | 117 |
| TABT | 2 | 7 |
| CONS | 28 | 63 |
| Total time within tasks | 148 | 208 |

[1] Minutes aggregated over decision function codes.

assigned at a time, i.e. one primary code and a secondary code. These codes can be interpreted in a manner chosen by a researcher. This would have allowed us to express both MCDM and MAPS in the same one-minute coding interval.

The coding and analysis related to the lag sequential analysis is also not without its problems. Comparing the information interpreted from the raw code presentation of lag sequential data analysis (LSA) (in Table 7.3) to the information interpreted from the aggregated codes of LSA (Table 7.4), we can definitely say that the Table 7.3 information is truer to the data, than are the interpreted results from the aggregated codes in Table 7.4. However, the interpreted results in Table 7.4 are simpler and more general than the raw results: does such a presentation connote more information? Sequence is a fundamental part of the raw data relationships in Table 7.3. Due to aggregation of codes to "simplify" the number of reported constructs, the temporality of the constructs (represented by raw data observations) at a detailed level within a particular task can be lost, although of course the generality of constructs increases.

Now that we knew that both analysis strategies have their advantages and disadvantages we were interested in comparing them in a systematic way. A side-by-side assessment of the analysis strategies can be performed using the ten features of analyzing relationships introduced in Chapter 4 (about social-behavioral research methods) and listed in the left-hand column of Table 7.5. The relation features are listed in order of "information gain" as presented by Brinberg and McGrath (1985). Remember that each relation between two variables *ij* is meant to correspond to the relation between two elements from the substantive domain (in this situation maps and decision tables) as well as the relation between two elements in the conceptual domain (in this situation two categories of decision aids, "map aids" and "MCDM decision aids"). That is, the relation between maps and decision tables should correspond to the relation between map aids and MCDM aids, and those two relations should correspond to the relation between counts of each aid as coded in the methodolgical domain. Because the count variables and the underlying data that we are using are the same in both the traditional and sequential analyses, i.e. our methodological domain, we can compare the two analyses in a systematic manner. To do this we simply step through the feature relations 1–10, i.e. compare the second and third columns in terms of their information gain for each analysis. The information in Table 7.5 is represented by entries in the cells of columns 2 and 3 of the respective feature in column 1. In stepping through the relations, we note that the traditional data analysis technique

*Table 7.5* Comparative relation feature analysis for research question 1.1 (task complexity influence on the various types of decision aids)

| Relation feature | General linear model | Lag sequential analysis |
|---|---|---|
| (relation between task complexity $i$ and decision aid $j$. Features 1–10 are described in Chapter 4) | (the factor is task complexity ($i$), dependent variable is map moves ($j$), covariate is MCDM moves ($g$)) | (task complexity ($i$) influence on decision aids ($j$)) |
| 1) presence or absence of a relation $ij$ | yes, factor 1 (low), task 2 (medium), task 3 (medium), task 4 (high) and MAPS move count and MCDM move count | yes, task complexity task 1 (low) and task 4 (high), and MAPS, MCDM, TABT, CONS minutes |
| 2) temporal order of a relation $ij$ | yes, a task # acts as control, data observed by task # | unknown |
| 3) logical order of a relation $ij$ | unknown | yes, task complexity is lower for task 1 than for task 4 |
| 4) direction of a functional relation $ij$ | yes, positive | unknown |
| 5) form of a functional relation $ij$ | MAPmove and MCDMmove are each a linear step function of task complexity | unknown |
| 6) deterministic or stochastic character of a relation $ij$ | unknown | unknown |
| 7) temporal stability of a relation $ij$ | unknown | unknown |
| 8) relation of element $g$ to element $i$ of the $ij$ relation | unknown | unknown |
| 9) relation of element $k$ to element $j$ of the $ij$ relation | unknown | unknown |
| 10) relation of element $l$ to both $i$ and $j$ or to the $ij$ relation itself | unknown | unknown |

(general linear model, GLM) provides more insight into the relationship between task complexity (*i* treatment) and decision aid (*j*) than does an exploratory sequential data analysis technique (lag sequential analysis, LSA), because the GLM column contains more information than does the LSA column. Four cells in the traditional analysis "general linear model" column contain information. "Unknown" constitutes no information. Only two cells in the sequential analysis column contain information. It is useful to point out that an entry of "unknown" in more basic cells, i.e. the features 1–3, actually jeopardizes what is known in the cells 4–10.

### 7.3.2 Research question 3.1

What is the relationship between the usage of maps appropriated in relation to decision models? To examine this research question we formulated the following null hypothesis: There is no relationship that exists between the number of map moves and the number of MCDM moves across decision sessions. We tested this hypothesis by relating the count of map moves and the count of MCDM moves using a Pearson correlation statistic. To do this we had to recode the counts as described for research question 1.1. The value of the Pearson correlation coefficient ($r = -0.124$, Sig. $= 0.001$) is statistically significant, but the correlation is very low. Although we reject the null hypothesis, the result indicates a very weak inverse relationship meaning that the map moves and MCDM moves are not likely to occur in a systematic manner across tasks.

When processing the data using a sequential analysis approach, in Table 7.6 we provide the count of minutes of use of decision aids for each of the three sessions. We note that session 3 (middle of the five sessions) has a much higher count of minutes that the other two sessions. Maps were used much more in session 3 than in session 1 or 5. MCDM tools were used about twice as much in session 3 as session 1. Table/text aids were hardly used at all. Consensus tools were used in session 1 about twice as much as in session 3 or 5. In using sequential data analysis we cannot really say how one set of aids was used in relation to the other aids, except to describe the

*Table 7.6* Total time spent with decision aids by session

| Decision aid category (DSCS) | Session 1 (minutes) | Session 3 (minutes) | Session 5 (minutes) | Total by aid |
|---|---|---|---|---|
| MAPS | 57 | 132 | 33 | 222 |
| MCDM | 106 | 198 | 159 | 463 |
| Table/Text | 19 | 27 | 9 | 55 |
| CONSensus | 48 | 19 | 19 | 86 |
| Total time in session | 230 | 376 | 220 | 826 |

relative amount of use. The answer for when they were used suffers again from the lack of information about simultaneity in the data coding.

*Discussion of findings and strategy assessment*    When performing a side-by-side assessment of the analysis strategies we see that there is as much information gain in the column with the traditional analysis performed with Pearson correlation as there is with the exploratory column under LSA (see Table 7.7). When we perform a correlation we know little about the temporal or logical relationship between MAPS and MCDM aids. There is nothing in the theoretical framework (for EAST2) that would help us, beyond the fact that maps and tables are both good conversation generators. In this side-by-side assessment there appears to be more con-founding relationships with other variables ($g$, $k$, $l$) that might influence the use of decision aids. In addition, variables representative of $g$, $k$, and $l$ might be controlled or measured, but just not participating in the analysis. For example, "$g$" could be representative of conflict, "$k$" could be representative of total years of group experience using GIS or decision tables, and "$l$" could be years of education in the group.

### 7.3.3  Research question 4.1

What is the relationship between map usage and decision functions, and between decision table usage and decision functions? To examine this research question for maps we formulated the following null hypothesis: No difference exists in map moves with respect to decision functions. We were interested to find out if the frequency of map appropriations differed with respect to the decision functions within the constraints of the prede-fined task. We tested the hypothesis using a one-way analysis of variance (ANOVA) procedure. The dependent variable was map moves and the factor was decision function. The decision functions included: 1) function structuring, 2) problem exploration, 3) criteria identification, valuation, and prioritization, and 4) evaluation and selection of alternatives. We also included the category *None* to represent the lack of activity. The differences in map move means were statistically significant for evalu-ation/selection of decision alternatives only, i.e. EA/SA in comparison to other functions (see Table 7.8). Thus, only for the EA/SA case can we reject the null hypothesis, suggesting that the mean of map moves in EA/SA is different from the other phases.

The finding highlights an observation that groups used maps predomi-nantly to visualize the evaluation results and much less to structure/design the decision problem. The question arises then: were the maps provided in Spatial Group Choice simply not adequate for problem exploration, cri-teria identification, valuation, and prioritization? Based on the analysis of variance, maps implemented in Spatial Group Choice played only a limited support role in the decision functions of the experiment. In an

*Table 7.7* Comparative relation feature analysis for research question 3.1 (usage of maps and decision models in relation to each other)

| Relation feature | Pearson correlation | Lag sequential analysis |
|---|---|---|
| (relation between usage of maps *i* and decision models *j*. Features 1–10 are described in Chapter 4) | (count of map moves (*i*) and the count of MCDM moves (*j*)) | (total time spent with decision aids (*i*) by session 1, 3, 5 (*j*)) |
| 1) presence or absence of a relation *ij* | yes, MAPS counts; MCDM counts | yes, session 1, 3, 5: minutes of MAPS, MCDM, TABT, CONS |
| 2) temporal order of a relation *ij* | unknown | unknown |
| 3) logical order of a relation *ij* | unknown | unknown |
| 4) direction of a functional relation *ij* | could be either positive or negative | 1 → 3 time increases 3 → 5 time decreases |
| 5) form of a functional relation *ij* | MCDM counts = f(MAPS counts) where MCDM counts is a linear function of MAPS counts | unknown |
| 6) deterministic or stochastic character of a relation *ij* | unknown | unknown |
| 7) temporal stability of a relation *ij* | unknown | unknown |
| 8) relation of element *g* to element *i* of the *ij* relation | *g* could be decision function | *g* could be decision function |
| 9) relation of element *k* to element *j* of the *ij* relation | *k* could be conflict intensity | *k* could be conflict intensity |
| 10) relation of element *l* to both *i* and *j* or to the *ij* relation itself | *l* could be attention loss as sessions progress; or *l* could be learning affect with same task; or *l* could be loss of motivation | *l* could be attention loss as sessions progress; or *l* could be learning affect with same task; or *l* could be loss of motivation |

*Table 7.8* ANOVA of map moves by decision functions. Only the significant differences are presented in the table

| (i) Decision function | (j) Decision function | Mean difference (i−j) | Standard error | Sig. |
|---|---|---|---|---|
| EA/SA | CI/CV/CP | 1.6486 | .227 | 0.000 |
| | FUNCS | 1.9730 | .227 | 0.000 |
| | NONE | 1.6157 | .228 | 0.000 |
| | PE | 1.6757 | .227 | 0.000 |

attempt to find an answer we examined the frequencies of map use by decision functions. Two decision functions in which group participants used maps most frequently were: 1) evaluation and selection of alternatives (EA/SA), and 2) identification, valuation and prioritization of criteria (CI/CV/CP). The three maps used most frequently were bar map representing attribute values, site situation map, and orthophoto image.

To further examine the use of maps by function we developed the following null hypothesis: No difference exists in the frequency of using map aids between session halves. We tested this hypothesis with paired-samples T-test statistical procedure. The results of T-test with first half mean = 0.62 and the second half mean = 1.84 (t = −4.570, Sig. = 0.000), allow us to reject the null hypothesis and indicate a different pattern of map use from MCDM use. Maps were used much more frequently during the second, more "integrative" half of experiment, than the first, so-called exploratory half, in all tasks.

To examine this research question for decision tables in a similar way to that of maps we formulated the following null hypothesis: No difference exists in the frequency of using MCDM aids between session halves. We expected to see less use of MCDM aids during the first half of the session than during the second half, since the first half of each experimental session was expected to involve more exploration discussion than integrative discussion. We used a paired-samples, T-test statistical procedure, to test the hypothesis. The results of T-test were not significant (t = − 0.427, Sig. = 0.670), thus accepting the null hypothesis and indicating a lack of difference in MCDM moves between the session halves. Additionally, the comparison of means of MCDM moves across the tasks confirmed that MCDM aids were used with similar frequency in both session halves.

When performing lag sequential analysis on the data we first examined the times within decision functions across the first, third and fifth sessions (Table 7.9). The complexity of tasks (simple, medium, complex) was randomized within each session. Some groups started task 1 in the first session, others started task 1 in the second sessions, others the third, fourth, and fifth. The reported sum of times will not equal the sum of times for the "complexity analysis", because the number of tasks included in this analysis is different from that compiled for complexity. We chose to examine sessions 1, 3 and 5 to get an idea of task performance over time.

We coded decision aid use during the different decision functions, and compiled the information according to session sequence, i.e. for sessions 1, 3 and 5, for all participants. We used the first (1), middle (3), and fifth (5 and last) sessions to compare the use of decision aids by each function in the decision functions coding system. The decision functions coding is listed in a normative order. Whether participants followed this order is not taken into consideration in this analysis. The rank order of the times from high to low are evaluation of alternatives (EA), criteria prioritization (CP), criteria identification (CI), select alternatives (SA), criteria valua-

*Table 7.9* Total time within respective decision functions for sessions 1, 3, and 5

| Decision function category (DFCS) | Session 1 (minutes) | Session 3 (minutes) | Session 5 (minutes) | Total by decision function |
|---|---|---|---|---|
| PE | 8 | 21 | 3 | 32 |
| CI | 32 | 55 | 28 | 115 |
| CV | 29 | 28 | 32 | 89 |
| CP | 41 | 95 | 61 | 197 |
| EA | 84 | 149 | 69 | 298 |
| SA | 36 | 28 | 27 | 91 |
| Total for all groups by sessions | 230 | 376 | 220 | |

tion (CV), and problem exploration (PE). Again, we point out the large session time for session 3 — perhaps explainable as participants really getting into the task, whereas session 1 was learning and session 5 was "let's get it over".

Now that we have some idea of the total time using maps or decision tables relative to each other according to decision function activity, we can compare the percentage of the time for MAPS use and MCDM use (see Table 7.10). During problem exploration the amount of time is approximately the same. However, for criteria identification, criteria valuation, and criteria prioritization there is definitely an emphasis on MCDM use, rather than MAPS. Once the criteria have been established, there is then an emphasis on MAPS relative to MCDM. Then, in selecting an alternative, emphasis changes back to MCDM. The sum of the percentages for MAPS and MCDM use do not total to 100%, because TABT and CONS make up the other portion of the time. The reader can see that in the "select alternatives" function, other decision aids account for 50% of the time, as might be expected since a consensus (CONS) capability is among the software to be used.

*Discussion of findings and strategy assessment*   The mean of map moves markedly differs between two session halves for each task. It was surprising to find that the participants used MCDM aids without much difference in the frequency of moves in both halves of the experiment. The prototype software *Spatial Group Choice* used during the experiment offered the participants both analytical functions useful for alternative evaluation and exploratory/visualization functions that could potentially be used during the problem exploration phase. We speculated that the first, more "exploratory" half of the experiment would be marked by more frequent use of maps than the second half — but only because of anecdotal evidence about maps as exploratory aids in various literature. The

*Table 7.10* Minutes of MAPS and MCDM decision aid use within decision functions in sessions 1, 3, and 5

| Decision function category (DFCS) | Session 1 (minutes) | | Session 3 (minutes) | | Session 5 (minutes) | | %MAPS minutes of function minutes | %MCDM minutes of function minutes |
|---|---|---|---|---|---|---|---|---|
| | MAP | MCDM | MAP | MCDM | MAP | MCDM | | |
| PE (Problem Exploration) | 6 | 2 | 6 | 14 | 1 | 2 | 40% (13/32)* | 56% (18/32)* |
| CI (Criteria Identification) | 2 | 27 | 30 | 25 | 0 | 28 | 28% (32/115)* | 70% (80/115)* |
| CV (Criteria Valuation) | 10 | 19 | 1 | 27 | 0 | 32 | 12% (11/89)* | 88% (78/89)* |
| CP (Criteria Prioritization) | 1 | 23 | 4 | 90 | 0 | 59 | 2.5% (5/198)* | 87% (172/198)* |
| EA (Evaluate Alternatives) | 33 | 28 | 90 | 29 | 30 | 22 | 51% (153/298)* | 27% (79/298)* |
| SA (Select Alternatives) | 5 | 7 | 1 | 13 | 2 | 16 | 9% (8/91)* | 40% (36/91)* |
| Total time for MCDM use | 57 | 110 | 132 | 190 | 33 | 158 | | |

* Total time for each decision function (i.e. denominator) taken from Table 7.9.

much less frequent use of maps during the first half of experimental sessions indicates the possible need for new types of maps, such as those for charting the lineage of collaborative discourse and problem structuration, as in a spatial understanding and decision support system (Jankowski and Stasik 1997, Moore 1997). However, it also points out the need for analytical, visualization aids that might help in exploratory decision situations.

A side-by-side comparison shows that traditional strategies (using ANOVA plus T-Test) and sequential analysis strategy (using LSA) have information gain for features 1 and 3, but not 2 and 4–10 (Table 7.11). There is likely to be a significant amount of confounding information through 8–10. Confounding information comes from variables that have not been considered in the analysis. Clearly, there is a broad-based set of information created, but it is not "very deep" in regards to what really influences what. Variables for feature relations 8, 9, and 10 could be considered, but were not in this analysis. These variables would be the next to be investigated, or least controlled for, while examining decision aid use.

### 7.3.4 Research question 4.3

Does task complexity influence group conflict? To examine this research question we formulated the following null hypothesis: No difference exists in the level of conflict between task 1 (least complex) and task 4 (most

*Table 7.11* Comparative relation feature analysis for research question 4.1 (relationship between map usage and decision functions, and between decision table usage and decision functions)

| Relation feature | Analysis of variance | Lag sequential analysis |
|---|---|---|
| (Relation between decision function $i$ and map usage $j$. Features 1–10 are described in Chapter 4) | (Factor is decision functions ($i$), including 1) function structuring, 2) problem exploration, 3) criteria identification, valuation and prioritization, and 4) evaluation and selection of alternatives; dependent variable map moves ($j$)) | (Decision function for sessions 1, 3 and 5 ($i$), minutes of MAPS and MCDM decision aid use ($j$) |
| 1) presence or absence of a relation $ij$ | yes, statistically significant EA/SA only yes, statistically significant for halves of session | yes, major variations |
| 2) temporal order of a relation $ij$ | unknown | unknown |
| 3) logical order of a relation $ij$ | yes, some decision function leads to decision aid use | yes, decision function leads to decision aid use |
| 4) direction of a functional relation $ij$ | unknown | unknown |
| 5) form of a functional relation $ij$ | unknown | unknown |
| 6) deterministic or stochastic character of a relation $ij$ | unknown | unknown |
| 7) temporal stability of a relation $ij$ | unknown | unknown |
| 8) relation of element $g$ to element $i$ of the $ij$ relation | $g$ could be another activity | $g$ could be function structuring |
| 9) relation of element $k$ to element $j$ of the $ij$ relation | $k$ could be another aid being observed | $k$ could be another aid being observed |
| 10) relation of element $l$ to both $i$ and $j$ or to the $ij$ relation itself | $l$ could be conflict intensity | $l$ could be conflict intensity |

complex). To test the hypothesis we used one-way analysis of variance (ANOVA), whereby the dependent variable was conflict level (measured by the length of time the groups were engaged in conflict situations) and the factor was task complexity (tasks 1 and 4). The results showed no statistically significant difference between the group conflict in tasks 1 and 4, hence we accept the null hypothesis.

A lag sequential analysis of the data set shows that there is not much difference in the percentage of time spent in group work with conflict between task 1 and task 4, i.e. compare column 3 (26%) and column 5 (37%) in Table 7.12.

*Discussion of findings and strategy assessment*   The finding from the traditional analysis that task complexity was not associated with the level of conflict between task 1 (simpler) and task 4 (more complicated) is somewhat contrary to current literature. Renn, Webler and Wiedemann (1995) present a diagram indicating that high levels of complexity are associated with higher levels of environmental conflict. However, their diagram also takes into consideration higher conflict being associated with a difference among three levels: knowledge level, experience and trust level, and worldviews and values level, respectively. Since all participants volunteered to take part in this experiment, and the experiment was situated in an environmental context, it is likely that most participants had a similar environmental perspective, even though we asked each participant to take on a stakeholder perspective. Perhaps this was the influence of a contrived, experimental setting versus a more realistic setting.

Although the results of the lag sequential analysis do not show a relationship between conflict and complexity, there is another interesting relationship that does appear in Table 7.12. When comparing the use of MAPS decision aids to MCDM decision aids within the category of group work with conflict, with the second control being task complexity, we see that in task 1 MAPS (11 minutes) and MCDM (10 minutes) aids are used about the same, whereas in task 4 the use of MCDM (52 minutes) aids is about nine times as much as MAPS (6 minutes) aids. Thus, decision tables coded as MCDM are not seen as useful when conflict or complexity is

*Table 7.12* Minutes using decision aids associated with group work and task complexity

| Decision aid category (DSCS) | Task complexity | | | |
| --- | --- | --- | --- | --- |
| | Task 1: simple | | Task 4: complex | |
| | GW without conflict (min[1]) | GW with conflict (min[1]) | GW without conflict (min[1]) | GW with conflict (min[1]) |
| MAPS | 46 | 11 | 14 | 6 |
| MCDM | 48 | 10 | 73 | 52 |
| TABT | 3 | 1 | 6 | 2 |
| CONS | 12 | 16 | 44 | 20 |
| Total time within Group Work (GW) | 109 (74%) | 38 (26%) | 137 (63%) | 80 (37%) |

[1] Minutes aggregated over group working relations codes.

viewed alone, but decision tables are rather useful when considering complexity and conflict together.

In the side-by-side assessment, the amount of information gain is "0", except for the additional insight that we gained by including decision aids in the LSA (Table 7.13).

*Table 7.13* Comparative relation feature analysis for research question 4.3 (task complexity influence group conflict)

| Relation feature | Analysis of variance | Lag sequential analysis |
|---|---|---|
| (Relation features 1–10 described in Chapter 4) | (Dependent variable is conflict level $(i)$ measured by the length of time the groups were engaged in conflict situations and the factor is task complexity $(j)$ — tasks 1 and 4) | (Percentage of time spent in group work with conflict $(i)$ between task complexity for task 1 and task $(j)$) |
| 1) presence or absence of a relation $ij$ | relationship not significant | no differentiation with $ij$ |
| 2) temporal order of a relation $ij$ | unknown | unknown |
| 3) logical order of a relation $ij$ | unknown | unknown |
| 4) direction of a functional relation $ij$ | unknown | unknown |
| 5) form of a functional relation $ij$ | unknown | unknown |
| 6) deterministic or stochastic character of a relation $ij$ | unknown | unknown |
| 7) temporal stability of a relation $ij$ | unknown | unknown |
| 8) relation of element $g$ to element $i$ of the $ij$ relation | unknown | unknown |
| 9) relation of element $k$ to element $j$ of the $ij$ relation | unknown | unknown |
| 10) relation of element $l$ to both $i$ and $j$ or to the $ij$ relation itself | unknown | element $l$ as decision aids might be related to the combined relationship between task complexity and conflict $(ij)$ |

## 7.4  Discussion of comparative assessment

Research questions and the anticipated findings from research encourage certain research designs. As part of research design discussed in Chapter 4, treatment mode selection and data strategy choice lead to analysis strategy choice — or at least we should say facilitate and/or constrain analysis strategies. Analysis strategies are what we employ to investigate the relations set out in research questions. The two strategies we used in this chapter (traditional statistical analysis and sequential analysis) are but only two strategies of a much larger set of choices presented in Chapter 4. Nonetheless, the two strategies are reasonably similar and different to encourage an insightful comparative assessment of analysis strategies.

Most social-behavioral researchers recognize that a particular analysis technique implements a particular type of relation between two or more variables. Thus, there are always constraints, some more fundamental than others. We decided to present the parametric statistical analysis as the traditional analysis strategy, since most researchers learned the techniques through graduate school training, and assumptions about their use are reasonably well known. The techniques are available within most general statistical software packages. However, the exploratory sequential strategy and its respective lag sequential analysis technique is a relatively newer approach to processing data. It is for that reason that we decided to introduce it second. Neither of the analysis strategies was useful in examining all of the questions. This dilemma encourages a mixed-method approach to data analysis in order to address all questions. This mixed-method activity encourages us to compare the analysis of social-behavioral relations in a unique way using Brinberg and McGrath's (1985) framework for feature relations.

Both types of strategies and their respective techniques make use of quantitative data as coded number of moves or number of time-cell counts. The traditional analysis was able to compare two sets of codes, i.e. MCDM and MAPS codes, within a given coding system, i.e. decision aid coding system, because we recoded the one-minute cells to two-minute cells to represent observations within a single coding system that occur simultaneously. The sequential analysis was better for comparing observations between the coding systems. The latter was not suitable for processing data within a single coding system because we did not recode it, nor did we take advantage of the primary and secondary codes within a time-cell as would have been possible.

Investigation of a social-behavioral relation between two (or more variables) can be differentiated in a number of different ways — in fact in at least ten fundamental ways as given by the relation features as listed in column 1 of Tables 7.5, 7.7, 7.11, and 7.13. The comparative assessment depicted in those tables shows that there are some fundamental differences in what might be expected to be very similar analyses, at least "on the surface". The ten relation features help to point out the differences between the traditional analysis and the sequential analysis. In Tables 7.5, 7.7, 7.11, and 7.13, the more the number of features resolved, the higher the information gain. Comparing and contrasting the entries in the tables allows us to show that there are differences in terms of information gain between the strategies.

Reviewing the specific results of our side-by-side assessment in terms of all four research questions for all relations, it is very interesting to note that not a single analysis in either traditional or sequential made it above feature 5. What this means is that we still know very little in regards to how decision aids are used in group decision making processes, at least in the context of this particular experiment. From another perspective it means that even a laboratory experiment is limited in how it can pre-establish social-behavioral relations in a laboratory setting. What it also implies is that we are still learning about the nature of "meaningful find-ings", since few have provided guidelines for what it means to establish meaningful findings. With such guidelines in place, it is much easier to understand how internal, external, construct, and statistical conclusion validity can be affected by the research design of a study. Remember that the research design includes specifying treatments, selecting a social-behavioral setting, designing data collection techniques, and anticipating data analysis to be performed.

Further detailed inspection of the side-by-side assessments for all four research questions show that the findings derived from a lag sequential analysis making use of aggregated data codes within treatments, i.e. codes analyzed by task and by session, do not preserve a "temporal order". That shortcoming is indicated by the cell entry labeled "unknown" in the right-hand column for the feature relation (2) in Tables 7.5, 7.7, 7.11 and 7.13. Based on those findings we strongly suggest that aggregating the code counts developed from a lag sequential analysis should be approached with caution in order to promote higher levels of information among treatment comparisons.

The treatments in this case were task complexity and sessions to help control the situations in which data about human-computer-human inter-action would be sampled. We thought at least that task complexity (as problem complexity) was a major underpinning to the need for different computer tools. We now know that adding conflict to complexity really does make a difference in decision aids that would be appropriated. Col-lecting (or at least coding) data with the intent of capturing "group atten-tion" is not as easy as it would seem. The coding has a lot to do with the construct validity. Although reliable codes can be collected, in terms of coders all agreeing that the same behavior is coded in a certain way, it is up to participants to establish the representativeness of those codes in terms of "meaningful behavior". Although we might claim construct valid-ity in this analysis, we would be hard pressed to say that internal validity has been achieved; that is, there may or may not be much evidence for causal relations in our findings. Causal evidence comes from findings that at least demonstrate temporal if not logical relations among variables. In addition, with the participants we have chosen to use in the experiment we could not claim much in the way of external validity, except to say that a few members of a community when sitting down with students, did indeed find that using GIS decision aids can readily help sort through complexity of habitat restoration site selection in a decision process.

## 7.5 Conclusion

In Chapter 4 we presented social-behavioral research methods for PGIS research to show that opportunities for mixed-method approaches to research are expanding. The analyses reported in this chapter point this out by elucidating how two different data strategies address the same set of research questions and experimental data. We reported on the use of three traditional statistical data analysis techniques — a general linear model, an analysis of variance, and a difference of means T-test — to examine research questions associated with human-computer-human interaction. We reported on the use of lag sequential analysis as a technique to use as part of an alternative analysis strategy. We showed that several of the questions are readily addressed by a traditional statistical strategy, as well as by exploratory sequential analysis. We also pointed out that some of the research questions are addressed by only one of the strategies, whereas four research questions are addressed by both. We suggest, however, that neither strategy represents the "end all approach in this analysis" — each has advantages and disadvantages. This is a point we made in Chapter 4, and have now demonstrated that point with more evidence.

Both strategies together constitute a mixed-method approach to analysis. Reporting on the combined set of findings, although useful, was not our goal in this chapter; we do, however, suspect that a rather interesting set of findings is likely to surface, since part of the findings are reported elsewhere in Jankowski and Nyerges (2001). In a similar, but different way, we chose not to focus on the fundamental differences between the strategies, as each of the strategies could be used to analyze data from only that perspective. Instead, we focused our analysis, hence relations between variables, on research questions that could be addressed by both strategies. Our choice was motivated by an interest in exploring the differences (similarities) in the analyses from a deeper perspective than simply reporting on empirical results in a study, although we did that too. By performing a comparative assessment of analysis strategies we wanted to go beyond just a presentation of mixed-method analysis results, and provide concrete evidence for why a mixed-method analysis is useful. That concrete evidence came in the form of the comparative assessment that shows the nature of the information gain for each analysis strategy, as we compare and contrast them. We were particularly interested in exploring and conveying to readers the nature of information gain from social-behavioral analysis, using a framework developed by Brinberg and McGrath (1985) that was introduced in Chapter 4. We know of no other such detailed analysis appearing in any research literature, although Brinberg and McGrath (1985) did provide a few brief examples.

The difference in the findings between the traditional analyses and the sequential analyses stems in part from the difference in data preparation for the analyses, and in part from how relations among data are preserved when a comparison is performed. In the traditional analysis we used the number of "moves", akin to the number of events, of a given coded observation. In the sequential analysis we used "time-cell counts" of codes, i.e. a count of the one-minute interval codes. The traditional analysis does not naturally preserve the sequence of moves, when performing analysis. That must be done with special effort by manipulating the data according to treatment control. Thus, using traditional statistical techniques to examine data that are sequential in nature has always been problematic. That difficulty is one of the major reasons why sequential analysis techniques were created.

Although sequential analysis is superior to traditional statistical analysis when working with sequential data, we have discovered that one can still face challenges with deriving findings. From our experience with code generalization for analysis, we conclude that aggregating the code counts developed from a lag sequential analysis should be approached with extreme caution. This conclusion is based on our discussion about the information preserved in a lag sequential analysis report of findings by "raw code sequence", in relation to the advantage of obtaining information about relationships through generalizing to supra-codes when we aggregate. Aggregation beyond a given level, i.e. by treatment or supra-code, can result in considerable loss of information. In the discussion of findings we noted that the feature relation for temporal order became unknown as we aggregated. What we choose as variables in the relation, based on the treatments we use as a foundation for analysis, is important. Thus, we have a significant trade-off when reporting the results of lag sequential analysis: simplicity with generality versus complexity with specifics. What is the "best information" — in the generality or in the specifics of the findings? A researcher's interest in findings based on an audience valuing the research (as discussed in Chapter 4) should help a researcher make the decision. Nonetheless, if one uses a data collection approach that preserves sequence, and a technique that can address that sequential relationship, we recommend that a researcher avoid the temptation to work the data in a way that does not preserve sequence, even though simplification of relationships at first looks very promising.

Choosing appropriate analysis techniques (as well as from among other than these) is based on what types of relationships are to be preserved and emphasized during analysis. Relationships in the substantive domain are the meaningful associations among actors and technology in specific decision making situations. Relationships in the conceptual domain are the meaningful associations among actors and technology as researchers conceptualize social-behavioral activity across a number of situations. The relationships in the methodological (analysis) domain attempt to work with data to make operational the conceptual relationships that in fact preserve the corresponding substantive relationships in the specific decision making situations being studied. Findings in a research study are about introducing new information where there might not have been any previously.

Our main concern at this time pertains to what we would need to do to be able to achieve higher levels of information gain than resulted from the analysis reported herein. Control treatments and better measurements are needed to lay the foundation for a more powerful analysis. We used task complexity as a "controlled treatment", but that was not enough to help qualify the character of the relationships. Measuring variables as another mode of treatment is essential. Laying out the data strategy in a more complete manner might have proved useful to make operational the relationships to be investigated; thus taking advantage of the control. After performing the assessment here we are more aware of the benefit of paying attention to the correspondences among the feature relations of the substantive, conceptual and methodological domains. With greater correspondence the findings will attain a higher level of information gain. The information gain can indicate our improved insight into the relationships among human-computer-human interaction variables about PGIS use.

The page space in this chapter allowed for the reporting of only one of the techniques for exploratory sequential data analysis, but there are other techniques that look similarly promising. Fisher cycles analysis uncovers the iterative character of codes (Sanderson and Fisher 1994). For example, in the decision functions coding system we suspect that various code sequences might appear more frequently than others, based on research reported by Poole and Roth (1989a, 1989b). In addition, transitions analysis can tell us the pair-wise sequencing of code types, i.e. the number of times one code type "transitions" to another code type. Obviously, Fisher cycles and transitions analyses are related to lag sequential analysis, but they uncover different information. A new release of the MacSHAPA software in 2000 should help us with this analysis, perhaps bringing in even more opportunities for analyzing the video-coded data we have on hand. As part of that analysis we hope to include a fourth coding system that focuses on "map use" — a coding system we have yet to apply to the data, given time and resource constraints. We expect to report on these analyses in the future. We encourage others to explore similar channels of social-behavioral analysis as we build toward a participatory geographic information science.

# References

Brinberg, D. and McGrath, J. (1985) *Validity and the Research Process*, Thousand Oaks, Sage.

DeSanctis, G. and Poole, M.S. (1994) "Capturing the complexity in advanced technology use: adaptive structuration theory", *Organization Science*, 5(2): 121–47.

Girden, E.R. (1992) *ANOVA: Repeated Measures*, Newbury Park, Sage.

Jankowski, P. and Nyerges, T.L. (2001) "GIS-supported collaborative decision making: results of an experiment", *Annals of the Association of American Geographers*, in press, March issue.

Jankowski, P., Nyerges, T.L., Smith, A., Moore, T.J. and Horvath, E. (1997) "Spatial Group Choice: a SDSS tool for collaborative spatial decision-making", *International Journal of Geographic Information Systems*, 11(6): 577–602.

Jankowski, P. and Stasik, M. (1997) "Design consideration for space and time distributed collaborative spatial decision making", *Journal of Geographic Information and Decision Analysis*, 1(1): 1–8;
    URL: http://publish.uwo.ca/~jmalczew/gida.htm.

Moore, T.J. (1997) "GIS, society, and decisions: a new direction with SUDSS in Command", *Proceedings, AutoCarto 13*, v. 5, American Society for Photogrammetry and Remote Sensing, Seattle, Washington: 254–66.

NOAA (National Oceanic and Atmospheric Administration) (1993) Technical notes from the Elliott Bay/Duwamish restoration program, April 14, 1993. *NOAA Restoration Program*, Sand Point Office, Seattle, WA.

Nyerges, T.L. (1991) "Analytical Map Use", *Cartography and Geographic Information Systems* (formerly *The American Cartographer*) for special issue on Analytical Cartography, 18(1): 11–22.

Nyerges, T., Moore, T.J., Montejano, R. and Compton, M. (1998) "Interaction coding systems for studying the use of groupware", *Journal of Human-Computer Interaction*, 13(2): 127–65.

Poole, M.S. and DeSanctis, G. (1990) "Understanding the use of group decision support systems: the theory of adaptive structuration", in J. Fulk and C. Steinfield (eds) *Organizations and Communication Technology*, Newbury Park, Sage, 173–93.

Poole, M.S. and Roth, J. (1989a) "Decision development in small groups IV: a typology of group decision paths", *Human Communication Research*, 15(3): 323–56.

Poole, M.S. and Roth, J. (1989b) "Decision development in small groups V: test of a contingency model", *Human Communication Research*, 15(4): 549–89.

Renn, O., Webler, T. and Wiedemann, P. (1995) *Fairness and Competence in Citizen Participation: Evaluating Models for Environmental Discourse*, Dordrecht, Kluwer Academic Publishers.

Rohrbaugh, J. (1989) "Demonstration experiments in field settings: accessing the process, not the outcome, of group decision support", in I. Benbasat (ed.) *The Information Systems Research Challenge: Experimental Research Methods*, Harvard Business School, MA: 113–30.

Sanderson, P.M. and Fisher, C. (1994) "Exploratory sequential data analysis: foundations", *Human-Computer Interaction*, 9: 251–317.

Sanderson, P., Scott, J., Johnston, T., Mainzer, J., Watanabe, L. and James, J.

(1994) "MacSHAPA and the enterprise of exploratory sequential data analysis (ESDA)", *International Journal of Human-Computer Studies*, 41: 633–81.

SPSS Base 8.0. (1998) *Applications Guide*, Chicago, SPSS Inc.

Todd, P. (1995) "Process tracing methods in the decision sciences", in T. Nyerges, D. Mark, M. Egenhofer and R. Laurini (eds) *Cognitive Aspects of Human-computer Interaction for Geographic Information Systems*, Dordrecht, Kluwer, September 1995: 77–95.

Vogel, D.R. (1993) "Electronic meeting support", *The Environmental Professional*, 15(2): 198–206.

# 8 Conclusions and prospects for future research

## Abstract

In this final chapter we summarize our conclusions about research findings concerning participatory geographic information use that we made in each of the separate chapters. We interpret the implications of the findings as contributions toward a participatory geographic information science. We discuss prospects for future research about a participatory, spatial decision making that makes use of geographic information systems by reflecting on the research framework we have utilized throughout the book. As a foundation of the framework used in this book, we emphasize a balance among theory, methods and substance in our studies about the use of participatory geographic information systems. We contend that the proposed framework and the studies constitute the basis of a participatory geographic information science. In this approach the theory guides the use of methods, which are applied to solve substantive decision problems involving locational (spatial) characteristics. The approach serves both the development of group decision support technology as in participatory GIS and research about the use of participatory GIS. The theory (Enhanced Adaptive Structuration Theory 2 (EAST2)) provides a conceptual map for understanding a group decision support situation, thus providing the basis for selecting appropriate methods and decision support tools for the participatory task at hand. EAST2 further provides the guidelines for empirical research investigations involving participatory GIS. The empirical investigations about the use of participatory, geographic decision support tools and methods in substantive decision situations, allow us to verify EAST2, enhance our understanding of the tools and methods, and in turn lead us to develop better methods of participatory GIS. The chapter concludes with a discussion of prospects for future research about participatory GIS use that can broaden and deepen the knowledge base associated with the still fledgling subfield of participatory geographic information science.

This chapter is comprised of two sections. In the first section, we summarize the conclusions about research findings that concern the use of participatory geographic information systems as a contribution to participatory geographic information science. The findings develop from a balance among three domains — theory, methods and substantive — through which a researcher can address research issues about participatory, spatial decision making. Since our findings concern only a limited number of studies and methods employed, in the second section we offer prospects for research about participatory, spatial decision making and the development of participatory GIS.

## 8.1 Summary of conclusions for research findings about PGIS use

This book started with the observation that most of the research concerning participatory, spatial decision making has been about GIS development rather than about GIS use, without a strong theoretical link between the two. This gap between the theory and applications created the need to develop an understanding of how GIS software with integrated decision support techniques is used in group decision processes, i.e. which components of computer technology fulfill decision support tasks and which do not. To provide a foundation for closing the gap we proposed an approach balancing theory, methods, and substance.

Beginning with a framework providing the bridge between a theory and applications of collaborative spatial decision making we formulated a macro-micro strategy for analysing decision situations. The macro-micro approach allows us to appreciate that every macro-phase in a macro strategy can have a different set of information needs, based on the collective needs of the micro-step activities. Consequently, a macro-micro decision strategy motivates (in large part) the requirements for decision support tools. Such information needs and the associated decision support tool requirements can only be addressed by a good understanding of the decision situation at the time and place (context) within which it occurs. The lack of such understanding has been the major stumbling block in group-based decision support, i.e. a flexible but thorough framework for unpacking the complexity of needs from a macro-micro perspective has not been proposed before. As an example of the macro-micro strategy for analysing participatory decision situations we used a matrix comprised of three columns representing macro phases: intelligence, design and choice, and four rows representing micro activities: gather, organize, select, review. The four micro activities together with three macro phases of the decision process constitute twelve "phase-activities" of the particular version of the macro-micro framework presented in the book. The significance of the labeling "phase-activity" is that a phase speaks to the issue of what is expected as an outcome in the overall strategy, while an activity is an action that takes place to facilitate creation of the outcome. The macro-micro strategy for analysing decision situations is a normative description of an expected decision process. In order to provide a more in-depth articulation of what can transpire during a participatory decision making process involving the use of GIS and other decision support technologies, we developed Enhanced Adaptive Structuration Theory 2 (EAST2). EAST2 is a network of constructs and their relationships providing a theoretical framework to organize and subsequently help explain a participatory decision process. As such, EAST2 has both a research and a practical value. From the research perspective, EAST2 helps to explain the expected and observed realizations of participatory decision processes involving inter- and intra-organizational groups and human-computer-human interaction. From the practical/application perspective EAST2 helps set up group decision support systems for specific decision situations. But how can a theoretical framework comprised of constructs and their relationships effectively contribute to building a participatory GIS?

Above we restated that most of the research concerning the use of GIS to support decision making has been about GIS development rather than about GIS use, without a strong theoretical link between the two. We also stated that the gap in understanding how GIS software, combined with other decision support tools, is used in group decision processes, which components of computer technology fulfill decision support tasks and which do not, could be closed by an appropriate theory. Such a theory would explain human-computer-human interaction in the context of participatory GIS rather than just describe it. Additionally, such a theory should be corroborated by empirical findings to see whether it provided a "useful organization" of ideas in the sense of illuminating the linkage between group decision support technology and its use. We argued that if there was a theory helping us to predict how (in what manner) groups will use computerized decision support tools in various participatory decision tasks, then we could propose more robust solutions for participatory GIS than the currently existing ones. We consider EAST2 to be such a theory. EAST2 provides the basis for developing participatory GIS and selecting tools appropriate for a given task due to its comprehensive character. It consists of a set of eight constructs, with 25 aspects as the basic *elements* of the theory that outline significant issues for characterizing group decision making, and a set of seven premises that describe the *relations* between the eight constructs. The 25 aspects in different combinations for each premise can "map" different relationships that may occur during a group decision making process involving human-computer-human interactions. The aspects in conjunction with the premises allow us not only to formulate research hypotheses about the use of participatory GIS and its likely outcomes, but also help us assess which methods and decision support tools will be likely to address decision support needs.

We demonstrated an application of user needs (task) analysis guided by EAST2 in Chapter 5. Task analysis of primary health care funding allocation in Idaho proceeded by examining convening, decision process, and decision outcome constructs of the framework of EAST2. The analysis of convening constructs allowed us to articulate what was important in setting up a decision task. The convening constructs included the identification of values, goals, objectives and criteria shared by the participants and the identification of decision support tools likely to benefit the collaborative effort. The analysis of process constructs allowed us to consider the dynamics of invoking decision aids and managing decision tasks. In the case of an analysed decision situation, two facilitated modes of participant interaction — private/public and public — were determined as feasible. The outcome constructs included the selection of comprehensive and efficient evaluation criteria, and the consensus-based ranking of Idaho counties on the basis of the need for primary health care services.

We used EAST2 framework again in Chapter 6, albeit in a different way than in Chapter 5, to guide us in the proposition analysis of a transportation improvement program in the central Puget Sound region. The propositions of EAST2 express relationships among the pairs of constructs. Analysing the relationships among constructs and construct-embedded variables of the transportation improvement process we were able to partially reconstruct needs, process, and outcomes in a multi-participant setting that highlight the state of PGIS use. The state of GIS use in this decision situation indicate there is much potential for decision support in the future. In fact, only two months after the first draft of this chapter was complete, the Puget Sound Regional Council requested proposals for the design and implementation on a WWW-based application for project transportation display and query (the co-authors did not apply).

In Chapter 4 we presented a variety of social-behavioral research strategies for conducting studies of PGIS use in part to show that opportunities for mixed-method approaches to research are expanding; but in Chapter 7 we demonstrated and then evaluated the use of mixed-methods. In Chapter 7, if we had used different research questions and different data strategies, it would have been easy to show that mixed-methods contribute to broader findings. Instead, we chose to demonstrate how two different data analysis strategies — traditional analysis techniques and lag sequential technique — that address the same set of research questions and experimental data can create different findings. Thus, we chose to mix the analysis methods, but control the research questions, rather than let the research questions and the analysis strategies vary. By performing a comparative assessment of analysis strategies using the correspondence feature relations introduced in Chapter 4, we wanted to go beyond just a presentation of mixed-method analysis results, and provide concrete evidence for why a mixed-method analysis is useful. Differences in the use of map aids and decision tables can be detected by both strategies. Task complexity did not matter much in differentiating the use of maps and/or decision aids. However, task complexity and group conflict together did make a difference in the amount of use of maps and decision tables — with decision tables being used when detailed analysis ranking was to be performed. As an assessment of the validity of those findings, we saw that information gain was different among the strategies. The traditional statistical techniques tended to have more correspondence among feature relations than did the lag sequential. However, the lag sequential could treat temporal considerations better because of the inherent component of time in the meaning of "sequential". Despite the usefulness of both strategies, we also concluded that neither strategy represents the "end all approach in this analysis": each has advantages and disadvantages. This is a point we made in Chapter 4 and then demonstrated in Chapter 7.

## 8.2 Prospects for future research about PGIS use in relation to current work

EAST2 is at this point more a theoretical framework rather than a fully verified theory. In order for it to become a theory capable of illuminating future spatial decision support systems for groups, it must be subjected to empirical verification dealing with diverse groups of participants representing different types of inter- and intra-organizational groups. We outlined different approaches to conducting empirical studies of groups in the context of participatory GIS in Chapter 4. Since it is neither practical nor feasible to analyze all possible combinations of group and spatial decision tasks an important question has to be asked. What are the conditions for the generalization of empirical research findings, so that a limited number of empirical studies can synergistically contribute to our understanding of the effective ways of employing computer technology to support participatory spatial decision making? We address this question as our next steps for future prospects of research. We do this by providing the reader with an idea of how one might chose complementary research strategies in a research agenda by focusing on the fundamental balance among the three social-behavioral research domains.

As we stated in Chapter 2, then again in Chapter 4, and demonstrated in Chapters 5–7, researchers' interests in emphasizing research domains determines the character of a research study. That is, a researcher's choice of conceptual, methodological, or substantive domain as first, second, or third choice dictates the emphasis of the domain in the study. Consequently, a different order for the three domains is the fundamental basis of a research strategy that results in a different type of research study (see Table 8.1). The choice of a lead domain establishes a research orientation. *Choosing the conceptual domain to lead* means your research is likely to be "basic research". Choosing the substantive domain to lead means that your research is likely to be "applied research". Choosing the methodological domain to lead means that your research is likely to be oriented to "methods research". However, it is very important to understand that a fully informed research study employs all three domains, as they are

*Table 8.1* Research studies about PGIS use: current work and future prospects

| Current studies | Research orientation | Main path | Domain pathway | Data strategy product | Analysis strategy product |
|---|---|---|---|---|---|
| Chapter 5 | applied | empirical | SMC | situation-driven observations | interpreted observations about participatory decision support capabilities useful for a decision situation involving funding allocation for primary health |
| Chapter 6 | applied | theoretical | SCM | situation-driven propositions | tested propositions about aspects of participatory decision support within a transportation improvement decision making process |
| Chapter 7 | basic | experimental | CMS | concept-driven design | tested hypotheses about human-computer-human interaction during a habitat restoration decision experiment |
| **Future prospects** | | | | | |
| Study X | basic | theoretical | CSM | concept-driven hypothesis | tested propositions describing different conceptual interpretations of a participatory decision support situation |
| Study Y | method | experimental | MCS | method-driven design | tested hypotheses about different approaches to gathering human-computer interaction data in experiments about participatory decision support |
| Study Z | method | empirical | MSC | method-driven observations | interpreted observations about the usefulness of task analysis in needs assessment for decision support |

Adapted from Brinberg and McGrath (1985), Table 3.1.

ordered by emphasis. We made this point earlier when we said that Chapters 5–7 are studies that use the domains in a different emphasis. At the same time, the choice of emphasis is what leads to choosing a particular research strategy composed of phases, i.e. research question articulation, treatment mode selection, data gathering strategy (setting and data collection), data analysis strategy, and reporting strategy.

In this book, each of Chapters 5–7 brings together the three research domains in different ways, resulting in different types of research studies. The different types of research studies are based primarily on different research strategies, i.e. different ways of asking research questions, different ways of gathering data, and different ways of analyzing data. Different research strategies are used as plans and then implemented as research studies in line with research orientations (see Table 8.1). Chapters 5 and 6 both have an applied research orientation because the leading emphasis is from the substantive domain, as indicated in the column labeled "domain pathway". Chapter 7 has a basic research orientation because the lead domain is "conceptual", as indicated by the "C" listed first under the column labeled "domain pathway". We designed our studies such that we could implement three different types of studies according to the combinations of research domains.

For the three studies reported here, our overall goal would be to develop findings that can be used to compare against current findings. Findings and hence potential knowledge about participatory GIS can be expanded upon for each of those three studies by choosing one of the other five domain pathways for each topic. Incremental finding expansion would occur if we were to choose the same lead domain, but an alternative second domain. For example, for the domain pathway in Chapter 5, which is SMC, we might choose SCM. In that case we change from an applied-empirical to an applied-theoretical strategy to find out more about the internals of the decision process at the Idaho Department of Health and Welfare. Alternatively, choosing a different pathway as in CMS would potentially provide findings from an experimental perspective, i.e. findings that provide specifics about human-computer-human interaction in decision processes that make use of decision support aids. The same kind of systematic consideration of research opportunities exists for the transportation and habitat topics as well.

Expanding further on the opportunities, from a combination of the three research domains — conceptual, methodological and substantive — there are actually six pathways for research studies in general. The other three pathways not covered in this book are listed as Studies X, Y and Z in Table 8.1. We see tremendous benefit in pursuing these other pathways in the future to enhance our perspective about researching PGIS use, when the research is grounded in a balance across the three domains. Despite the prospects of the additional three pathways that have not yet been tapped, the prospect for research studies is actually much greater than indicated by six pathways. The 18 research strategies described in Chapter 4 indicate that there are actually many more choices for designing studies than the six pathways. For each of the pathways, there are choices for each of the research phases: research question articulation, treatment mode selection, data gathering strategy (setting and data collection), data analysis strategy, and reporting strategy. However, what Table 8.1 shows is that the prospects for research studies can be laid out systematically. A systematic research agenda can be constructed for any given substantive topic, theoretical conceptualization, and methodological approach, as researchers make choices based on their interests. What at one time seemed like a jumble of research topics, methods, designs, theories, etc. can be interrelated in various ways to promote a well-informed research agenda. It is up to researchers to choose wisely in relation to the six pathways and 18 strategies. Using the ideas of three orientations, three paths, and three domains that compose six pathways, considerable flexibility exists to compare research findings: the essence of knowledge building. In a systematic way we expand on the prospects for research rather than narrow the prospects. Consequently, prospects for research about participatory geographic information systems use as a contribution to participatory geographic information science appear enormous.

Those prospects will continue to grow, as we better understand how to undertake meaningful research in the light of the proliferation of technology in both professional and everyday life. More meaningful research can be undertaken when a researcher is more informed about the relationships among the three research domains, including the myriad of choices possible among components for research strategies. As technology proliferates, knowing about the influences and impacts of PGIS use in society can help encourage/discourage technological proliferation in certain ways.

Changes in the way we interact are being encouraged/discouraged by access to changes in the way we use data communications technologies, e.g. the internet, for which the volume of data on phone lines has now passed voice transmission. Of course, data communications is only one of the large number of technologies, but in terms of participatory geographic information science it is a major aspect of technological change. Since the opportunities for technological change are enormous, the opportunities grow for research directed toward a better understanding of how this change influences human-computer-human as well as computer-human-computer interaction in society. Some people have said modern communications technologies have decreased the size of the world. However, what that really means is that the geographic space of the world, at multiple scales in any given day, is becoming a much more intimate part of everyone's life. What people do to each other and their environment is becoming a more sensitive concern as the impacts of humans on environment and environment on humans are more closely intertwined than ever before. PGIS use will only increase in importance as wireless data communications technology proliferate to the extent that millions and millions of people in the future will be making use of "GIS in the phone", as computers miniaturize and become more mobile.

We hope the material in this book assists others in their research pursuits to examine PGIS use in society, as it has helped us thus far. As geographic information technology continues to permeate all aspects of life, we hope that this book will assist others gain a better sense of the craft of research from a scientific perspective, particularly in relation to contributions toward a participatory geographic information science as one way to enhance our knowledge about our world.

## Reference

Brinberg, D. and McGrath, J. (1985) *Validity and the Research Process*, Thousand Oaks, Sage.

# Index

For Product Safety Concerns and Information please contact our EU
representative GPSR@taylorandfrancis.com
Taylor & Francis Verlag GmbH, Kaufingerstraße 24, 80331 München, Germany